# ASSEMBLING JAPAN

# ASSEMBLING JAPAN

## Modernity, Technology and Global Culture

Edited by
Griseldis Kirsch, Dolores P. Martinez and Merry White

PETER LANG
Oxford • Bern • Berlin • Bruxelles • Frankfurt am Main • New York • Wien

Bibliographic information published by Die Deutsche Nationalbibliothek.
Die Deutsche Nationalbibliothek lists this publication in the Deutsche Nationalbibliografie; detailed bibliographic data is available on the Internet at http://dnb.d-nb.de.

A catalogue record for this book is available from the British Library.

Library of Congress Cataloging-in-Publication Data

Assembling Japan : modernity, technology and global culture / [edited by] Griseldis Kirsch, Dolores P. Martinez and Merry White
pages cm
Includes bibliographical references and index.
ISBN 978-3-03-431830-3 (alk. paper)
1. Japan--Civilization--1868- 2. Japan--Civilization--Western influences. 3. Culture diffusion--Japan. 4. Culture and globalization--Japan. 5. Technological innovations--Japan. I. Kirsch, Griseldis, 1975- II. Martinez, D. P. (Dolores P.), 1957- III. White, Merry, 1941-
DS822.5.A87 2015
952.03--dc23
2015002065



ISBN 978-3-0343-1830-3 (print)
ISBN 978-3-0353-0767-2 (eBook)


Hochfeldstrasse 32, CH-3012 Bern, Switzerland
info@peterlang.com, www.peterlang.com, www.peterlang.net



This publication has been peer reviewed.

Printed in Germany

# Contents

# Acknowledgements

The editors would like to thank two anonymous donors, the Faculty of Languages and Cultures, SOAS and Boston University for supporting the publication of this volume. We would especially like to thank our very patient contributors as well as Laurel Plapp at Peter Lang.

*A note on the text.* Japanese names in the text are in western order: first name followed by surname. Except for words now commonly used in English, all Japanese terms have been romanized according to the Modified Hepburn System.

GRISELDIS KIRSCH AND DOLORES P. MARTINEZ

# Japan as an Assemblage[1]

## The Quandary

What does it mean to assemble and to be assembled? Here we use the terms to describe Japan as a modern nation that both makes and represents itself, while it is also imagined, scrutinized, studied and represented by non-Japanese. In the global culture of late modernity this is the condition in which all nation-states find themselves, but Japan appears to hold a distinctive position in this system. Attempts to understand Japan's place in global modernity often describe Japan as reluctantly and only partially 'international', and frequently have posed three questions: to what extent has the world become a global village (cf. McLuhan 1967); to what degree has Japan become a worldwide presence (Matray 2001, Preston 2000); and what international forces are at play in Japan (Hook 2012)? In contrast, the essays collected in this volume aim to reconceptualize Japan as fundamentally global, seeing this state as the result of long-term processes that have influenced contemporary social discourses about and within the nation. By describing the complexity of ways in which Japan is unique as well as completely enmeshed in the more generic processes of modernity, our goal is both to produce a robust cross-cultural analysis and offer a nuanced understanding of the historical and cultural forces within the

1 We must thank the anonymous readers whose comments helped us hone our argument, as well as David Gellner and Andreas Kirsch for their patient re-readings. Our co-editor, Merry White, read this chapter various times and gave us final comments, some of which have found their way into the text. Any errors, however, remain our own.

Japanese nation-state. To achieve this, the additional question of how best to understand the antimony between being local and unique as well as global and similar to everyone else needs to be considered.

Recent anthropological theorizing has tended to reject theories of a homogenous global culture and its smooth progression post-Cold War (westernization, the triumph of capitalism, etc.), countering them with concepts such as that of hybridity (Bhabha 1994), fusion, friction (Tsing 2005), disjuncture (Appadurai 1996) and, more recently, assemblage (Collier and Ong 2005). These contemporary theories try to account for the fact that shared economic and political structures, as well as a shared reliance on modern technologies, have *not* resulted in producing identical nation-states. For this reason, modified versions of Wallerstein's (1974) centre and periphery model have been re-examined to explain the continued under-development of many of the world's southern nation-states – but how to explain the powerful discourse about the supposed distinctiveness of (over)-developed Japan?[2] One way to understand this representation is to examine the role played by the narratives known as *nihonjinron* (theories about the Japanese) in constructing Japan's idiosyncratic character. Here we outline a history for these socially constructed narratives in order to illustrate how assemblages come into being and are continuously reworked, acquiring powerful ideological strength.

'Japan is unique, and cannot be compared to any other nation in the entire world'. Such statements have been made within and outside of Japan in the mass media since the end of the Second World War. They can be traced further back, but gained particular currency during the era in which its economy expanded on the back of technological innovation (1974–91), leading to the growth in Japan's economic power and the acknowledgement that as a nation-state it was also part of global capitalism. *Nihonjinron* arguments implied that if Japan was going to be part of a largely western phenomenon, then it would be a singular component of the world's system. No matter that its corporate capitalism resembled Germany's (Dore 2000);

2 Harvey (2005) is one of the more interesting attempts to get to grips with this uneven spread of development.

that its system of company consultation had been created by an American (Aguayo 1991); that its democratic constitution was mostly written by foreigners; its education system was modelled on that of the United States; or that its post-war industrialization had been achieved on the back of nuclear power – it was claimed that Japan's ability to forge these elements into an economy that became 'number 1' (Vogel 1979) was due to its 'unique' past and cultural qualities.

*Nihonjinron* as a discursive set of theories still exists in the twenty-first century and in all its forms it emphasizes Japanese cultural, physical, psychological and social extraordinariness. Such theorizing ends by creating distance between Japan and the countries to which it is being compared, presenting both sides as irreconcilably different. Despite their many flawed assumptions and sometimes ludicrous claims – there exist books on the unique Japanese brain, nose and digestive system – *nihonjinron* became a significant rhetoric in post-war Japan and to some extent they have hindered inter-cultural exchange with the rest of the world (Befu 2001, Yoshino 1992). Yet for any scholar of nationalism statements of Japanese uniqueness are not necessarily hostile 'to any mode of analysis' as Peter Dale (1986: 100) has argued. Nation-states perforce develop discourses of group distinctiveness that call upon long histories, links to the soil and arcane customs, as well as a shared language and a collective identity (cf. Anderson 1991, Gellner 1983, Hobsbawn and Ranger 1983). However, from a western perspective, Japan may not only seem 'different', but also rather exotic in the most Orientalist and travelogue of ways: it has samurai, Zen and geisha as well as *manga*, anime and sushi!

Frequently cited as one source of this 'different' difference, is the era of Japan's closure 'to the world' (*sakoku* 1633–1853), implying that such long term insularity was bound to breed exceptions and exceptionality. Pre-war nativist theories about the Japanese climate, its 'unique' seasons, the threat of living with typhoons and earthquakes or the experience of being a nation of islands (see Watsuji 1961) have also been used to construct theories of why the Japanese are *so* different. Moreover for countries in which belonging is premised on a narrative of immigration such as Australia and the United States, Japanese myths of autochthonous origins and its post-war restraints on immigration, resulted in closing the doors to any sort of sharing of experience in the process of nation building.

National constructions of history and uniqueness notwithstanding, it is also important to note that Japan has had a long history of interaction and exchange with others. The country's most common exchanges have been with its East Asian neighbours: with China from the second century CE onwards (cf. Tsunoda 1964); and with Korea, which frequently acted as mediator for Chinese culture. For nearly 1,500 years, Japan paid tribute to Imperial China, adopting and adapting aspects of Chinese culture and society over the centuries it traded with the middle kingdom (*Chūgoku*). From the sixteenth century it also established trade relations with the Portuguese, and later the Dutch, while extending its trade routes into Southeast Asia. During this era Japan also expanded into Ezo (Hokkaido) and the Ryūkyū Islands (Okinawa). Thus the *sakoku* era was not a time of total isolation: Korean and Chinese diplomats regularly visited Edo, and the shogunate maintained its trade routes throughout East and Southeast Asia as well as continuing to trade with the Dutch.

From the Meiji Era (1868–1912) onwards, western countries, such as Britain, the United States and Prussia/Germany established economic and political relationships with Japan. In both the pre- and post-Meiji eras ideas and technologies were imported, improved, adapted and also discarded (Tobin 1992). In short, Japan has long been part of a wider world system. Moreover during Japan's Imperial era (1868–1945), the extension of a form of Japaneseness was promulgated through its colonial institutions in Hokkaido, Okinawa, Manchuria, Korea, Taiwan and other parts of its Pacific empire (Saaler and Koschmann 2007).

Accordingly throughout its history 'Others' (Outsiders/foreigners/immigrants) were not conceived of as entirely different human species, but were readily embraced and incorporated as Japan interacted with others and later expanded its empire. The immediate post-war era also saw Japan's successful incorporation of many aspects of life associated with modernity: democracy, capitalism and the continued adaptation of, as well as further innovations in, western science and technology. However, as noted above, after the Japanese Empire was dismantled, theories of Japanese uniqueness began to gain currency. There had been nativist agitators in the past who had called upon Japan's history and tradition as markers of difference from East Asian others and who had argued that this distinction allowed Japan to *lead* these others (McNally 2005). Such assumptions of difference and superiority had been central to Japan's colonial expansion, but the discourse altered as Japan worked to become a nation-state after 1945. It no longer sought to be inclusive of its non-Japanese subjects but became about excluding all non-Japanese. It became a *nationalist* discourse; this is an important shift that needs some elucidation.

In the 1980s and early 1990s *nihonjinron* called upon the early successes of modernization – while at the same time postulating that all foreign ideas had been indigenized, becoming 'truly Japanese', and held them up as examples of Japan's unique ability to succeed on the West's terms. It was this sense of having managed to keep up with the western Joneses that had underpinned Japan's colonial expanses in the early half of the twentieth century and which had led to a sense of superiority towards the countries it invaded and colonized. Post-war, much as Japan adopted western institutions and material culture, it nonetheless considered itself to be part of Asia, and still a superior power (Iwabuchi 2002). However Japan's colonial and military actions during the first half of the twentieth century, the apparent reluctance (and sometimes outright refusal) to deal with this past, as well as the nation's alliance with the western bloc during the Cold War, prevented Japan from completely interacting with Asia throughout most of the post-war period. As during the Meiji Era, the West was once again the focus of Japanese attention.

This is the context within which the emergence of post-war *nihonjin* theories must be understood. Effectively blocked from looking towards the 'East', Japan looked towards the West. Furthermore, having national symbols that were regarded as tied to its expansionist and militarist past by its close neighbours, made it difficult to find a positively connoted post-war, post-military national identity amidst considerable American influence during the Occupation (1945–52) and after. Moreover because Japanese culture was seen as having survived much change more or less intact, theorizing on what made Japan special as well as different from a seemingly overwhelming West, made sense in a way that it provided people with a distinctiveness they could embrace.

Throughout the 1980s and during the time of the bubble economy (1986–91), then, publications of *nihonjinron* treatises boomed and also became increasingly popular in the West, where they were used to explain why Japan, a small island nation with few resources other than that of its large population, had become economically successful. Attempts to internationalize Japan, and to thus make it more open to investors who had been frightened away by a strong currency and an apparently impenetrable culture (Katō 1992) were only half-hearted – and directed towards predominantly English-speaking countries. Thus even as it continued to be fully involved in a process of economic and political globalization, post-Imperial Japan was never seen as truly international or 'global' in the same way that Britain once had been or the United States seemed to be. Indeed, academic work on global Japan has examined the nation-state both in terms of its economic reach (Wan 2001) and its soft power around the world,[3] but it rarely has addressed the question of whether or not Japanese culture and society had in any way become more inclusive. Rather there have been frequent publications on Japan's resistance to immigration and its less than successful attempts to become 'more international' (cf. Chung 2010). The Japanese themselves seemed to take pride their reputation as impenetrable, ferocious samurai-like business competitors and inheritors of a rich cultural tradition that foreigners could only ever admire but never really understand. To reiterate, Japan has developed a nationalist discourse that downplays its long-term engagements with the rest of the world.

Yet through the 1990s Japan remained a sort of centre state for other Asian countries, skewing centre–periphery models that placed the West at the core of the globalization process. In fact the same criticism was initially made of Japan's centrality that has been made about 'western-style' globalization, namely that it was only a one-way exchange which replicated colonialism rather than being a balanced system of give and take. Moeran (2000) has even argued that Asia was seeing a kind of *Corollanization* by Japan, with its industrial and cultural products flooding Asian markets. Japanese

3 The body of literature on soft power and Japan is vast, but see McGray (2009) and Otmazgin (2014) as examples.

politicians and intellectuals talked about a 'return to Asia', a re-integration of Japan into the Asian (economic) context, which was welcomed by some Asian countries, such as Malaysia, that were keen on Japanese investment; and eyed distrustfully by others, such as China and South Korea, who feared a return to Japanese imperialism, albeit through economic and cultural rather than militaristic means. This re-entry into Asia was seen as an example of Japan's soft power, an interesting downplaying of a policy that was built on the 'hard' economic influx of Japanese products that included everything from toilets to automobiles, as well as ground-breaking new forms of technology.

As Japan slid into recession from 1991, it re-discovered its Asian neighbours culturally as well as economically. In economic terms, an increased number of goods began to flow – not just from Japan to the Asian mainland but also in reverse. Culturally, however, it took a little longer for the flow to become stable, particularly in the nations where memories of Japan's past aggression were still strong.

Since the new millennium the interest in Japanese products has grown in Asia – and both material as well as cultural products from other Asian countries have gained more of a foothold in Japan. Culturally, this tendency has been accompanied by a small but decisive 'Asia boom' within Japanese popular culture. Other Asian countries, which had been missing from the conceptual maps of Japanese popular culture, returned to the attention of producers and consumers (Schilling 1999, Iwabuchi 2002, Kirsch 2015). More recently, the earthquake and Fukushima nuclear disaster in March 2011, resulting in the decision to close down all of its nuclear power plants in the immediate aftermath of the disaster and again for maintenance in 2013, mean that Japan now relies on other countries for most of its energy needs. Additionally Japan has also become much more involved in economic and political activities on the Asian mainland, especially in China and South Korea, despite persisting territorial disputes.

Hence what has been going on in Japan in its relation to the outside world in the past few decades challenges the common stereotype that globalization inevitably means westernization, although it firmly places Japan very much within the flows of the post-war new world system. We use the term 'system' to suggest that the modern condition is one of intricate interconnectivity, which may well be skewed toward core states such as the United States, the European Union member states and now China, who exploit the cheap labour force and resources of peripheral regions. Since a system is a form of complex relationships, this means that the flows can never be entirely in one direction. The rise of many Asian countries, with Japan having led the way, demonstrably makes the point that the world is not so easily categorized into centre and periphery as it was in the 1970s and 1980s: there are complex processes taking place. Many peripheral regions now claim their share of the centre – or are becoming regional centres themselves. In short, the processes of globalization are complex and irreducible to a simplistic or singular model.

Thus, Japan has always been an interesting case – which may have fostered the assumption by the *nihonjinron* literature that Japan was 'special'. Definitely a core state when it came to economics, it also was not seen to be western and hence not in the same league as the United States and Europe. Accordingly its cultural delimitation as found in the *nihonjinron* literature becomes understandable – though its worst manifestations are not necessarily excusable. Moreover Japan's cultural products, which have been arriving on western shores in various waves since 1853 (Napier 2007), have come to occupy a much more central status recently: Japanese *manga* and anime are now widely distributed and consumed around the globe (Allison 2006, Condry 2013, Yano 2013). Japan itself continues with its unabated consumption of cultural goods from other countries not just western ones. Thus old *nihonjinron* tropes of Japan versus the West no longer made sense. Products, both material and cultural, from other Asian countries were and are consumed in Japan, although they may have initially only occupied niches among the population, such as in the case of Japanese fandom for Hong Kong Chinese stars (Iwabuchi 2002) and later with the *hanliu* (Korean Wave) across Asia (e.g. Mōri 2004, Iwabuchi and Chuah 2008). These Asian cultural products were and still are consumed alongside domestic and 'western' ones, underlining once more the complexity of the system and the difficulties one faces when trying to talk about globalization.

## Thinking 'through' Assemblages

Taking into account the fact that globalization's complex processes involve long-term developments offers possibilities for a different sort of analysis of 'global' Japan. First, the very concept of globalization always leads to a re-evaluation of the self vis-à-vis the other or, in the words of Robertson (1995: 40), globalization invokes 'the creation and incorporation of locality, processes which themselves largely shape, in turn, the compression of the world as a whole'. Localization, re-nationalization as well as conflicts within, and across, regions are a common feature of globalization, while trade and cultural exchange continue, albeit in unbalanced and sometimes unexpected ways (Appadurai 1996). Within the complex flows that dictate our world today inevitably come interactions we term cultural globalization. The academic discourse on cultural globalization (e.g. Pieterse 2003, Hopper 2007) is very well developed, but many have focused on how American culture, or its soft power, has dominated, dividing the world (again) into core and periphery states. In short, one is not modernized, or even westernized, but Americanized in such models. Japan might well seem a good example of this but, as we have argued above, its involvement in the process of cultural globalization could be said to have been ongoing for two millennia and the results have been anything but a Chinese pre-Meiji Japan or an American post-war state.

Thus modernization, its technology and the processes of globalization all form a complex field: one of tensions, of interesting dynamisms and synergies. These are the focus of this volume, which aims to contribute to the broader discussion on Japanese global flows in chapters that examine the range of cross-cultural consumption through the lens of Japan's interaction with the global community. As already noted, rather than critiquing *nihonjinron* and its problems (cf. Refsing and Goodman 1992), the chapters in this volume aim to go beyond the usual tropes of Japan versus the West, and so come to paint a different picture of Japan. The Japan of this volume is one that interacts with other Asian countries while continuing to interact with the West. The approaches here are manifold – some papers examine the topics of modernity, technology and Japan's global experience though popular culture (B. White, Martinez, Kirsch, Horne and Manzenreiter), others through food (M. White), travel (Yano), economics (Wong), cultural politics (Hendry) and technological innovation (Katsuno) – but all of them aim to show how this interaction takes place and how it is structured.

It is worth making some distinctions at this point. As noted above, hybridity, and more recently assemblage, terms that have been applied to such cultural and social developments, both have been critiqued for the implication that the resultant object, in this case the nation-state, is somehow inauthentic or 'invented'. For example, as a modern society, some might say that Japan is no longer what it was; it is no longer even 'oriental' but somehow western. Or in a broader sense: if Japan, like other nation-states, has invented traditions and censored/edited its history in order to make a case for its unique identity, is Japaneseness authentic? Such suppositions ignore the sociological view that all reality is socially constructed (Berger and Luckman 1967) and fall back on Platonic notions that somewhere there is the real world or a real Japan whose ontology we can study and understand. What is really, authentically, purely Japanese? These are questions to be asked of all nationalistic ideologies and the critiques made about *nihonjinron* being a discourse of exclusion applies to all such philosophies. We prefer to see authenticity as a concept in which the definers, their intentions and the contexts surrounding the 'authentic' are the objects of inquiry rather than ascribing objective validity to the term.

To put it in another way: all modern societies are hybrid or assemblages, they are all the result of particular conjunctures (Sahlins 1981), disjunctures (Appadurai 1996) and it is these that form the field of study for any anthropologist. This was the tack taken at the time the panel on which this volume is based was held in Oslo, Norway in 2007. In order to come to an anthropological understanding of modern Japan, we argued that scholars also needed to consider longer term formations, the *longue durée*, which underpinned the creation of the twenty-first-century nation-state we all studied. These processes, as outlined above, included a recent history of interaction (both aggressive and more peaceful) of Japan with Asia and the West, but this interaction also had longer roots that we as anthropologists could not consider, relying instead on the work of a new generation of historians of Japan who are challenging *nihonjinron* assertions (e.g. Gordon 2013). By limiting the discussion to the modern or postmodern era, we required our contributors to examine Japan in light of not just multiple and rapid social changes, but also within the context of an era that saw rapid advances in technology.

Such an approach would seem to favour Collier and Ong's (2005) use of Latour's (2005) conceptualization of the assemblage and re-assemblage of the social, referring to his focus on the reproduction of social structures as related series. We argue that *nihonjinron*, although not structures per se, form an important part of this process (as outlined in the previous section) since their claims to Japan's distinctiveness support a discourse of national authenticity and allow for powerful, political ways of organizing and reorganizing Japanese society. Theorizing about assembling or reassembling the social in this abstract as well as concrete fashion is useful in understanding the development of Japanese modernization in the Meiji era in general. As touched on above, it was during this point in time that the oligarchs of the new empire looked towards the West and made any number of choices: French, German and British advisors for the navy and army; Scottish engineers to teach in Japan and help build its railway system; German doctors and scholars to teach in medical schools and advise on a new constitution, etc. After the war, it could be argued, the choices were not solely that of the Japanese, but of its Allied Occupiers: the state would become a constitutional monarchy with democratically elected political leaders. However, this element of decisions taken and choices made mitigates, in our view, against the more Latourian approach to assemblage in which the relationships amongst humans, animals and technology appear to offer no such 'free' choice: we are not really actors, but actants enmeshed within and acted on by our social networks. As actants we can only play out social narratives that insist on individuality while configuring us as types: male, female, young, old, married, single and, pace Marshall McLuhan, we are also constituted by the technology around us. Choice would seem to drop out of such a model and humanity's imaginative and creative potential is lost within the institutional and structural assemblages into which we are born.

This view, one which resonates with concepts such as Marx's dominant ideology and Gramsci's hegemony, appear to leave little room for any ethnographic study other than that of social models and institutions – something the anthropology of Japan has specialized in since Nakane's seminal *Japanese Society* (1970). Latour salvages his position by saying that such an approach becomes the study of the links and processes of assemblage and, yet, even within this re-assemblage there can be no methodological individualism, only actants. While this seems a logical critique of the assumption that the individual has total free will, unconstrained by the social in any way whatsoever; it also is a problematic counter-assumption that begs the question: which is the chicken, which the egg? If humans make the social and the social makes humans – how is it that change ever occurs?

In answer to this the Japanese case becomes very instructive. Considering it as an example of a conjuncture, as Sahlins would have it, in which a nineteenth-century capitalist West encountered the feudal East, it was a meeting that ended rather differently than the encounter of China or Korea with the West. Japan's leaders literally reassembled the country rather than outright rejecting or resisting the West, but did so precisely in order to oppose the minority position to which the western powers wanted to relegate the new state and did this while apparently conforming to dominant western models of what it meant to be 'civilized'. Consequently, both the creation of new national structures and the (re)emergence of *nihonjinron* can be analysed through Deleuze and Guattari's conceptualization of assemblage as an act of becoming which is 'a capturing, a possession, a plus-value, but never a reproduction or an imitation' (1986: 13). For these theorists, assemblages require both a deterritorialization and dismantling of the various components that are about to be put together. Deterritorialization leads to unexpected metamorphoses and dismantling leads to an analytical stance. Both involve a taking apart of constitutive elements, allowing those who are doing the disassembling and re-assembling to take a political position; the processes afford a possible resistance to the dominant Other and can be revolutionary. Consider how in its post-war guise, Japan has continued to assemble itself in such a way that it remains both itself and a modern nation-state, acting on an international level, contributing to and partaking in something that can be termed global culture.

Assemblages are not just, as Deleuze and Guattari would have it, chaotic and unpredictable (thus resembling both conjunctures and their resultant disjunctures), but also essentially pro-creative. The pun is intended, the nation-state as an assemblage is not just new or unique onto itself; it also seeks to endlessly reproduce itself both biologically and socially through maintaining and populating its institutions. The nation-state reproduces also through the force of desire – not sexual in this case, but desire experienced as nationalist sentiment. Desire is paradoxically a socially constructed – as well as a productive – force that can engender radical change rather than just faithful reproduction.

In taking the position that modern Japan is an assemblage, as are all nation-states, this volume and its chapters aim to challenge the dominant view of Japan as a society that is impossible or difficult to comprehend, à la *nihonjinron*, to one that can be apprehended in terms of both political desire and individual choices that forge a national identity from a minority position: deterritorializing what it needs in terms of knowledge, technology and political ideology – in short, dismantling what it takes from others – in order to (re)assemble Japan.

Without having asked our contributors to adopt a Deleuzian perspective, they each have grasped the implicit task of all scholarly endeavour: in their desire to understand something (not everything) about Japan and the Japanese, they have dismantled their ethnographic data, what they have amassed as students of that society, and then reassembled it in a paper for this volume. Merry White, Yano, Hendry and Martinez, for example, all examine aspects of Japanese experience that seem ubiquitous throughout global modernity: coffee, air travel, the rise of indigenous movements and film. Each, however, looks at these shared facets of modern life somewhat differently. White describes how, as with so many adopted and adapted products in Japan, coffee has acquired its own way (*dō*), but has done so over many decades that involved similar processes as in the European case – Japanese coffee houses in the early twentieth century were places for political and cultural discussion, indicating a global engagement, and not just the province of haute *kohii*. Yano considers how air travel not only joined Japan to the West and the rest, but also provided a tool for already interested and keen young Japanese women to become more fully involved in the global system. Through the narratives of the first Japanese stewardesses for Pan Am, she traces a process of knowledge transfer that could be said to have begun in the Meiji Era. Hendry examines a somewhat similar, if more complex, case of global engagement by considering how the vanishing original inhabitants of Japan, the Ainu, and the proud inhabitants of an ancient independent kingdom, the Okinawans, connected with a global indigenous peoples' movement and experienced a cultural resurgence. Martinez focuses on the difficultly in pinning down what is 'Japanese' in a Japanese film industry that shares its technology and some of its techniques with every other film-making nation on the planet, asking if the very concept of 'Japaneseness' intervenes and becomes a discourse that adds to persisting *nihonjinron* versions of Japan's identity.

Kirsch, Bruce White, Manzenreiter and Horne examine equally global phenomena: television, music, football (soccer) and fandom, but trace different aspects of Japan's participation in these mass media. Kirsch considers the renewed relations with mainland Asia: Japan's 'rediscovery' of its Asian Others, specifically, China. In particular she considers how one of the chief agents in perpetuating notions of Japanese homogeneity – Japanese television – suddenly discovered Otherness. Bruce White takes another stance on this, as he looks at the fans of two musical genres in Japan: Reggae and Hawaiian. As with Merry White's work on coffee, Bruce White traces a commitment to this new form of music that is also political and generational. Its fans see themselves in opposition to 'mainstream' Japanese, having other values, values reflected in the political messages of Reggae and surfer culture's more laid-back approach to life. They are Japanese who, like Yano's stewardesses, see themselves as both very much of their society while also being open to experiencing whatever the global flow might bring to their shore. For Manzenreiter and Horne, football provides another example of a global conjuncture with unexpected consequences in the Japanese case. The Japanese want to love and succeed in the global spectacle that is football, but the battle for hearts and minds remains very much a local, political and corporate-led encounter.

The fans in all these examples, however, participate in a wider sphere than just of Japan and this aspect of what has been termed 'soft' power is interesting to consider. Can we really separate it from the realm of economic hard power? In all the articles discussed so far, the economic has played a role in the dissemination of a new lived experience: coffee is part of a global import and export business; air travel is intrinsic to the modern commerce; film and television may have their artistic and aesthetic values, but, as with music, also need to be profitable; while some would argue that football has been corrupted by its increasing immersion in the global financial system. The difficulty in separating the economic and other forms of hard power (the military and its technology) from soft or cultural power would point to one more reason why global flows never progress smoothly. Ironically, in the Japanese case the concern is often raised with the possibility of profound cultural change: is Japan really too western? Are we in danger of turning Japanese?

Wong and Yau consider a more limited case study of Japan's counter flow: how its business practices have fared in Hong Kong, a situation in which certain concepts seen to be Japanese and forming good business practice have been used to help form a new middle-class identity. Again, the result is one of unexpected conjunctures and assemblages. Hence, one way in which we can reconceptualize the global situation is as one in which every flow can result in difference rather than increased homogenization. Sahlins' concepts are crucial to Wong's and Yau's analysis and could be seen to form the basis for Katsuno's work on Japanese nation branding and robotics. 'Robot' may be a Czech, word, but automata have existed in various societies from ancient times. Japan had its own *karakuri ningyō* (automata), but more than that, it had and still has very different concepts of what constitutes an animate object. As Foster (2008) has shown, in Japan *anything* is capable of acquiring an animating spirit given enough time, love and care. Thus the Japanese attitude towards robotics and robots seems different from that of the 'Christian' West. This, however, does not mean that the world of amateur robotics is vastly different from that in other parts of the world – geeks, *otaku* and obsessive scientists are basis of a flow of technology that is perhaps understudied.

In trying to understand the assemblage that is Japan the nation-state we need to, as Sugimoto (2010) has argued, always ask: which Japan? What these various chapters do is make the reasons for such a question clear. When we talk of a unique Japan with its closed culture do we mean the Japan of young Japanese women, or of working class men? Is it the Japan of middle-class housewives and their white-collared husbands? Are we talking about the country that dreams of international stardom for its national team, as well as of coffee aficionados whose global connections really matter in that search for the perfect cup? And what does the existence of fandom indicate – since fandom so often cuts across assumptions of cultural incomprehension? 'The Japanese' is a large generalization that negates the existence of Ainu and Okinawan activists, artists, anti-nuclear protestors, NGO volunteers, military personnel, the differently abled, those of other sexualities, the elderly and the homeless. In short, it denies class, gender, regional, occupational and ideological difference in a nation-state of 124 million people. It also ignores social change in each sector or niche over time. Simultaneously it builds a nation, an imagined community of 'we' Japanese and foreign Others, assembling Japan through a dismantling of its subjects and their subjectivities.

None of the chapters in this volume perpetuate *nihonjinron* discourses or aim to revive them, but they all ask questions about what being 'Japanese' means in the wider world, how this complex identity is being negotiated by all the actors involved and what influence the 'Other' has in these assemblages. Each chapter assembles and disassembles Japan, examining different aspects of society and its interaction with 'different' agents, sometimes having contrasting results, sometimes coming to similar conclusions. Different Japans emerge in each chapter, leaving each reader to assemble their own conceptions of Japan.

## References


Aguayo, R. (1991). *Dr Deming: The Man Who Taught the Japanese about Quality*. London: Mercury.

Allison, A. (2006). *Millennial Monsters: Japanese Toys and the Global Imagination*. Berkeley: University of California Press.

Anderson, B. (1991). *Imagined Communities: Reflections on the Origin and Spread of Nationalism*. London: Verso.

Appadurai, A. (1996). *Modernity at Large: Cultural Dimensions of Globalization*. Minneapolis: University of Minnesota Press.

Befu, H. (2001). *Hegemony of Homogeneity: An Anthropological Analysis of Nihonjinron*. Melbourne: TransPacific Press.

Berger, P. L. and Luckmann, T. (eds) (1967). *The Social Construction of Reality*. New York: Anchor Books.

Bhabha, H. (1994). *The Location of Culture*. London: Routledge.

Chung, Erin Aeran (2012). *Immigration and Citizenship in Japan*. New York: Cambridge University Press.

Collier, S. J. and Ong, A. (eds) (2005). *Global Assemblages: Technology, Politics and Ethics as Anthropological Problems*. Oxford: Blackwell.

Condry, I. (2013). *The Soul of Anime: Collaborative Creativity and Japan's Media Success Story*. Oxford: Blackwell.

Dale, P. (1995). *The Myth of Japanese Uniqueness*. London: Routledge.

Deleuze, G. and Guattari, F. (1986). *Kafka, Toward a Minor Literature*. Translated by R. Bensmaïa. Minneapolis: University of Minnesota Press.

Dore, R. P. (2000). *Stock Market Capitalism: Welfare Capitalism: Japan and Germany versus the Anglo-Saxons*. Oxford: Oxford University Press.

Foster, M. D. (2008). *Pandemonium and Parade, Japanese Monsters and the Culture of Yōkai*. Berkeley: University of California Press.

Gellner, E. (1983). *Nations and Nationalism*. Oxford: Blackwell.

Goodman, R. and Refsing, K. (eds) (1992). *Ideology and Practice in Modern Japan*. London: Routledge.

Gordon, A. (2013). *A Modern History of Japan: From Tokugawa Times to the Present*. Oxford: Oxford University Press.

Harvey, D. (2005). *Spaces of Neoliberalization: Towards a Theory of Uneven Geographical Development*. Stuttgart: Franz Steiner Verlag.

Hobsbawm, E. and Ranger, T. (eds) (1983). *The Invention of Tradition*. Cambridge: Cambridge University Press.

Hook, G. D. (2012). *Japan's International Relations: Politics, Economics and Security*. London: Routledge.

Hopper, P. (2007). *Understanding Cultural Globalization*. Cambridge: Polity Press.

Iwabuchi, K. (2002). *Recentering Globalization; Popular Culture and Japanese Transnationalism*. Durham, NC: Duke University Press.

—— and Huat, C.B. (eds) (2008). *East Asian Pop Culture. Analysing the Korean Wave*. Hong Kong: Hong Kong University Press.

Katō, S. (1992). 'The Internationalization of Japan'. In G. D. Hook and M. Weiner (eds), *The Internationalization of Japan*, pp. 310–16. London: Routledge.

Kirsch, G. (2015). *Contemporary Sino-Japanese Relations on Screen. A History, 1989–2005*. SOAS Studies in Modern and Contemporary Japan. London: Bloomsbury.

Latour, B. (2005). *Reassembling the Social, an Introduction to Actor-Network Theory*. Oxford: Oxford University Press.

McGray, D. (2009). 'Japan's Gross National Cool'. <http://foreignpolicy.com/2009/11/11/japans-gross-national-cool/> accessed June 2015.

McLuhan, M. (1967). *The Medium is the Message: An Inventory of Effects*. New York: Random House.

McNally, M. (2005). *Proving the Way: Conflict and Practice in the History of Japanese Nativism*. Cambridge, MA: Harvard University Asia Center.

Matray, J. I. (2001). *Japan's Emergence as a Global Power*. Westport, CT: Greenwood Press.

Moeran, B. (2000). 'Commodities, Cultures and Japan's Corollanization of Asia'. In M. Söderberg and I. Reader (eds), *Japanese Influences and Presences in Asia*, pp. 25–50. Richmond: Curzon.

Mōri, Y. (ed.) (2004). *Nissiki Kanryū. 'Fuyu no Sonata' to Nikkan Taishū Bunka no Gendai'*. Tokyo: Serika Shobō.

Nakane, C. (1970). *Japanese Society*. London: Pelican.

Napier, S. (2007). *From Impressionism to Anime: Japan as Fantasy and Fan Cult in the Mind of the West*. New York: Palgrave.

Otmazgin, N. K. (2014). *Regionalizing Culture: the Political Economy of Japanese Popular Culture in Asia*. Honolulu: University of Hawai'i Press.

Pieterse, J. N. (2003). *Globalisation and Culture, Global Mélange*. Lanham: Rowman and Littlefield.

Preston, P. W. (2000). *Understanding Modern Japan: a Political Economy of Development, Culture and Global Power*. London: Sage.

Saaler, S. and Koschmann, J.V. (eds) (2007). *Pan-Asianism in Modern Japanese History: Colonialism, Regionalism and Borders*. New York: Routledge.

Sahlins, M. (1981). *Historical Metaphors and Mythical Realities, Structure in the Early History of the Sandwich Islands Kingdom*. Ann Arbor: University of Michigan Press.
Schilling, M. (1999). *Contemporary Japanese Film*. New York: Weatherhill.
Sugimoto, Y. (2010). *An Introduction to Japanese Society*. Cambridge: Cambridge University Press.
Tobin, J. J. (ed.) (1992). *Re-made in Japan: Everyday Life and Consumer Taste in a Changing Society*. New Haven, CT: Yale University Press.
Tsing, A. L. (2005). *Friction: An Ethnography of Global Connection*. Princeton, NJ: Princeton University Press.
Tsunoda, R., de Bary, W.T. and Keene, D. (compliers) (1964). *Sources of Japanese Tradition*. New York: Columbia University Press.
Vogel, E. F. (1979). *Japan as Number One: Lessons for America*. Cambridge, MA: Harvard University Press.
Wallerstein, I. (1974). *The Modern World-System*. New York: Academic Press.
Wan, M. (2001). *Japan between Asia and the West: Economic Power and Strategic Balance*. Armonk, NY: M.E. Sharpe.
Watsuji, T. (1961). *A Climate: A Philosopical Study*, trans. by G. Bownas. Tokyu: print. Bureau, Japanese Govt.
Yano, C. (2013). *Pink Globalization: Hello Kitty's Trek across the Pacific*. Durham, NC: Duke University Press.
Yoshino, K. (1992). *Cultural Nationalism in Contemporary Japan: A Sociological Enquiry*. London: Routledge.

# PART I

# Roots and Branches

MERRY WHITE

# Café Society in Japan: Global Coffee and Urban Space

## Origins

In Japan coffee is normal; the most common of social beverages – more so than tea. It is of the wider world as well as one of the earliest globalizing commodities and currently trades second only to petroleum. From its first uses in North Africa to the present, it has been on the move and quickly has become a local beverage. However, coffee has always had to travel considerable distances to become an ordinary part of life for most of us: the major consuming countries are not, except for Brazil, major producing countries. Coffee's markets are diverse and continue to grow as consumers are educated to want more choices and to respond to trends in taste and place.

Japanese engagement in this market flow of beans and the cultures of coffee taste has changed over time: its 'globalization' in Japan has in part related to the fact that the beans travel and in part related to the influences of western coffee drinking. In a most interesting twist, it was Japanese production and entrepreneurship, some argue, that gave the impetus to the worldwide coffee market a century ago. Currently we see coffee made in Japanese coffee equipment (most made by the Hario and Kalita companies) all over the world, used with the techniques that these tools demand. Its 'Japaneseness' has become a brand. Japan's beans, like those of specialty coffee menus everywhere, come from Brazil, Ethiopia, Colombia and elsewhere. Japan's quality testers (QT certified) are finicky and certify a varietal or a roast with fastidious certainty. These blends are often superior: while a blend in most of the rest of the world is less trusted, seen to be created to hide the taste of lesser beans, in Japan and in specialty coffee shops elsewhere, they are created by a master who knows how to make a brew of complementary beans, a drink better than the sum of its parts. Consumption is about trust.

Coffee has given character to and inspired connoisseurship of several kinds in social spaces in Europe and Asia while also being the catalyst – no more than a cup at a table needed – for a wide range of engagements. In the earliest locations of social coffee drinking, the cafés of the eastern Mediterranean, in Turkey and points east and south, men sit at tables in town squares talking for hours over the cups that come unbidden as soon as one sits down. Coffee bears witness to every type of discussion, confession, challenge and repartee. Coffee offers rest and stimulation, silence and community. The American politician on the stump knows to stop in at a town's hub, its coffee shop, for a 'cuppa joe' and a donut, but more than that, for the contact with the central social organization beyond the town hall, the church and the school. The invitation to share coffee goes well beyond the beverage itself. 'Let's have coffee' has different tonalities depending on the space chosen and on the uses of the moment for which coffee gives ample excuse.

Coffee and its cafés have also provided the loci for artistic and political movements. In Paris, Michelet said that coffee is 'the great event which created new customs and even modified the human temperament' (Barthes 1987: 190). In France again it was said that the salon stood for privilege, but the café for equality. This 'equality', often a challenge to established power, has sometimes given the space of the café a bad name. As one observer noted: 'If a café has a police record, it is an interesting place.'

Thus coffee has had its enemies. In seventeenth-century England, shortly after coffee became popular there, women campaigned against the brew because it was said to diminish men's virility. In Germany Frederick the Great forbade the roasting of coffee – seen as fuel for sedition – and he hired sniffers (*Kaffeeschnüffler*) to search out coffee roasting in unlicensed establishments. Coffee was the stuff of revolution in America also, as anti-British sentiment drove people to drink coffee instead of tea, and to leave tea to the British whose colonies amply provided them with their imperial drink. The coffeehouses of my political youth on the US East

Coast in the 1970s were also informal headquarters for anti-establishment demonstration planning.[1]

## Coffee in New Japanese Spaces

I focus here on the role of coffee in creating new kinds of places within metropolitan spaces and settings for the development of 'urbanity' in Japan, most particularly through the coffeehouse or café (*kōhii hausu* or *kissaten*). While a village market square maintains the relationships and character of a face-to-face community, the city café, with open access, encourages new connections as well as anonymous, temporary contiguity. The global and local influence each other in such urban places, but do not necessarily interact more deeply.

Since the first appearance of a modern cafe in Tokyo in April of 1888, cafés have been places where people have gathered to be both together and alone, for social recreation or for being private in public. Learning the new urbanity of the Meiji Era (1868–1912) was a project in which people became modern and democratic, learning the responsibilities of citizens, adopting a set of ideas, behaviours, structures and laws. When cafés were classrooms for modernity, they encouraged patrons (and do sometimes today) to learn to be 'urbane', to learn new options for dress and behaviour and to learn what at the time was becoming a form of civil society. This new urbanity was/is different from the Edo (1603–1868) experiences of neighbourhood, class and social engagement. It is part street wisdom, part cultural and political awareness and part what Elias (2000) calls moral character: the proprieties and the values that support or insist on it are the

1 While much of this material appears in a book I published in 2013, *Coffee Life in Japan*, this essay, which was prepared before the book, does not directly repeat the material of the book. Moreover, this version contains newer, updated material, gathered since its first iteration and newer than what appears in the book as well.

products of the contacts, movements and settlements of people in such places of witness and exchange.

In such places, urbanity means not the sophisticated glamour or doubtful morals of the *flaneur* or *boulevardier* in a transient, placeless culture, but what might be called instead 'globalism' – an interaction between likely and unlikely people and places, where an absence of the strictures of roles in the predictable institutions of family, school and workplace, allow for experimentation and choices in a changing society. The Japanese coffeehouse, novel in the early twentieth century, was seen as *the* symbol of the modern. According to Tipton (2000) the café was more significant than the Japanese Diet (Parliament), as it was the locus of expression of a popular will. Habermas' (1991) view of the English coffee house of the eighteenth century as a locus of class mingling and democratic communication is a possible parallel to this view.

Taking a seat in a café did not automatically introduce a visitor to modernity or to a new community, though it might educate him or her to its possibilities. People who came to live or visit Tokyo from the hinterlands would observe people whom they might desire to emulate, but the boundaries between their cultural niches could prevent direct intercourse. So a café might become a replica of a village for some, allowing predictable encounters with known acquaintances, but for many urban Japanese the flexibility and anonymity of these new urban settings, or 'communities of choice', were more attractive than the propriety-maintaining village square.

Both coffee and the places – social or personal – where it is consumed have consequently become quintessentially Japanese. In the initial wave of coffee influence Japanese learned about the beverage and its uses from early European visitors to Japan. Coffee was introduced first in port cities such as Nagasaki, in the seventeenth century – at about the same time that it became popular in western Europe – where Portuguese missionaries and traders, as well as the Dutch, initiated the Japanese into the ways of coffee. Initially it was used medicinally, as noted by von Sieboldt, a German botanist and doctor who visited Japan in the 1830s, and it was drunk as a pleasurable stimulant (with the addition of honey) among the prostitutes of Nagasaki who found it useful, keeping them alert to the customers who might otherwise cheat them if they fell asleep. In 1804, an essayist named

Oota wrote that he had coffee at the Nagasaki magistrate's office: 'I had something called kahii – something to do with powdered black roasted beans … It tastes burned. I can't stand its taste' (Terada 1933: 74). If one were over-stimulated one might eat an *umeboshi* (a pickled apricot-plum) as antidote.

Not until the Meiji Era did it become a generally available social beverage, its consumption a sign of urbanity. The first significant import was in 1877 when 18 tonnes arrived in Japan from Brazil. Japanese migrant workers in Brazil were the instigators of coffee importation. The entrepreneur who helped to create Japan's leading position in coffee consumption and taste was a Brazilian returnee (*nikkeijin*): Ryo Mizuno, Japan's first coffee tsar. At the turn of the twentieth century, Mizuno created the market for coffee, the flow of the commodity, and set the scene for the cafés such as his own Café Paulista, which expanded the taste for the drink.

The place where one sipped coffee became as important as the drink itself. As noted above, in the rapid modernization of the Meiji Era coffee was a symbol of modernity, and even more so were the places where it was drunk. These coffeehouses and cafés, begun in imitation of European and Brazilian styles, became 'local' and modern as Japanese social spaces.

The first café of note was created by the son of a Chinese translator for the Foreign Ministry, Ei-Kei Tei, as his Chinese name is transliterated into Japanese (Hoshida 1988). Tei's coffee history begins and ends in America. He was born in Nagasaki, but his father, a Chinese translator for the Foreign Ministry and ambitious for his son, sent him to Yale University – considering that as an international youth in Japan he might do better with good English and a foreign degree. In New Haven in the 1870s, however, he developed a taste for coffeehouse life, journeying to New York for this decadent indulgence. Tei appears not to have done well in school because he was sickly as well as because he was having rather too good a time. Having left without a degree, he took a slow route home to Japan by way of England and the rest of Europe, and much impressed with the London version of the coffeehouse, he returned to establish his own in Tokyo's Ueno Hirokoji district, an entertainment quarter.

Tei's coffeehouse, the Kāhiikan, was an instant success when it opened in April of 1888. Apparently the original kanji used for the name signified 'for better or for worse', in a playful and stylish sense. A very masculine club-like place, filled with stuffed leather furniture, writing desks, the newspapers of the day on racks, billiard tables, resting rooms and bathing facilities, it gave the new middle class man and the recently de-classed samurai a milieu for testing modern urban identities. The luxurious amenities were soon exploited beyond the meagre cost of a cup of coffee and keeping up the premises drove Tei into debt, eventually forcing him to close the house. With other family tragedies to endure, he became seriously depressed. A friend saved him from suicide, found him travel papers under a Japanese name (Tsurukichi Nishimura) and sent him to Seattle, then a prospering port town, where much of America's coffee came by ship. Here the trail runs thin, but there is some evidence that he worked in a mercantile establishment where he sold coffee, descending into dishwashing and then unemployment. He died young, at 39, and his gravestone is in a Seattle cemetery, visited by relatives and people in the Japanese coffee industry. Cafés, however, went on to prosper in Japan, to proliferate, diversify and regionalize.

By the end of the Meiji Era and the beginning of the Taishō (1912–26), cafés took on many styles, from the English-style Kāhiikan to Mizuno's Brazilian Café Paulista, which welcomed the Ginza demimonde in 1907. The branching of coffee houses from this period led to the cabaret on the one hand, serving alcohol and boasting elegant, trend-setting, eroticized, scandal-focusing *jokyū* (waitresses); and to the *jun kissa* on the other, the 'pure café' where no music, booze or women distracted writers (like Kafū Nagai) from their intellectual engagement and counter-establishment politics – the home of a rarefied aesthetic for the new urbanity.

The cafés of this period were globalizers as well, as Donald Richie noted, the cafés were 'where one first glimpses the foreign innovation that will shortly become Japanese' (Richie 2002: 52). Contemporary fashions and personal style were both on show. The café waitresses themselves became characters, figuratively and literally, in a new public theatre. Whereas geisha in the past had been trend-setters in hair, clothing and conversational novelties, now *waitresses* were expected to represent the avant-garde in fashion, and had to spend heavily on the new fashions, going into debt to pay for western hairstyles as well as creative new patterns and modes of wearing kimono (Dalby 2001).

Other novelties included food. Western dishes were first introduced to the public in the café – where dishes like gratin, spaghetti, melted cheese toast and pilaf appeared, and later, the caramel pudding called *purin* (Cwiertka 2006). While these foods were not considered meals, only snacks, they made inroads in taste and created a demand for products such as tomato ketchup, cheese and pasta, until then not used in the home kitchen. In addition, *karē raisu* (curry rice) now known to all Japanese as an ordinary food and a staple for industrial and military menus was introduced in coffeehouses in Tokyo, much as it had been as a colonial food in British seafront coffeehouses at least one hundred years previously.

Another novelty in the Taishō and early Shōwa (1926–89) periods was the café as music venue. The *meikyoku kissa* (famous music cafés) were places where, before most people had phonograph players, classical music was – and still is – reverentially played. Jazz cafés were also extremely popular. The historian Ishige (2000) argues that jazz, which he calls an intellectual music, went best with coffee, a beverage leading to 'dry inebriation' while blues and other popular music accompanied the 'wet inebriation' of alcohol. Cafés were also the first places where unescorted young modern men and women (*mobo* and *moga* respectively) could meet socially with relative impunity.

Organizers and proponents of new social and political thinking would hold meetings in cafés where the social formlessness of the space admitted all. Feminist discussion and the creation of a significant feminist movement in this period took place in cafés where strategic discussions could be publicly private, circumventing the Peace Preservation Law which forbade meeting in large groups. Undaunted by restrictions on the size of gatherings women organized simultaneous multi-sited meetings in several cafés near each other, where two or three women would meet for ostensibly genteel discussion. One would act as a runner between designated cafés to maintain communication. Marxists, Trotskyites and home-grown political movements also had homes in certain cafés, but the participants would move just ahead of the watchful police.

In the late 1960s and early 1970s student coffeehouses had become staging areas, planning sites for demonstrations, as well as para-medical stations between encounters with the police. Some cafés near Tokyo University in Hongo became headquarters for the student take-over of a building and nowadays veterans of the student movement return there for nostalgic reunions.

To give an ethnographic example of the variety of café experiences: in 1963, I was taken late at night to a coffeehouse in Tokyo where customers disrobed completely, had their bodies painted with large soft brushes in a deep blue, and then were pressed against the walls hung with sheets, leaving blue imprints of their bodies. I did not realize until a visit to the Centre Pompidou in Paris in 2007 that what I experienced in 1963 had been an evocation of the Kleinian *Anthropometrie* – in which naked female models were painted with Klein's signature cobalt blue paint and similarly pressed against blank sheets of paper. The café in Tokyo on the first anniversary of Klein's death in 1963 was paying homage to a man who, ten years before, in 1953, had electrified the Tokyo art scene. At this event there was indeed coffee, but it was merely the hospitable requirement, not the centre of attention.

Coffee houses have always had different personalities by city and region. During the Taishō era Tokyo's cafés were the first stop for visiting overseas artists, political thinkers and writers, and the launching pad for those internationalizing Japanese headed for the Asian mainland or Europe. The movements of such people included travel to similar café cultures in Shanghai, Berlin, Paris and London. At least 60 per cent of Tokyo-based artists in the late Taishō and early Shōwa eras studied and worked in Europe. Their experiences overseas made them consider the café in Japan as an extension of their travels abroad, almost as expatriate locations at home.

While Tokyo was seen as the centre of a western-leaning literary and political avant-garde, Kyoto in the Taishō era attracted Japanese intellectuals, writers and aesthetes. Kyoto evinced a different form of urbanity from that of Tokyo – as an older capitol during an era of high court culture, it had its own *iki* (flair) and it did not experience the earthquake of 1923, nor was it significantly damaged in the Second World War. The city remains a repository of older forms of cafés as well as a home to newer styles. The former include the Café Tsukiji of quintessentially European decor and menu, serving small nut tortes and Viennese coffee, playing Mozart on a gramophone. This was where the novelist Junichiro Tanizaki's coterie met, smoking and sipping on the padded red velvet seats. Tsukiji is today an example of what are called sepia cafés, places whose nostalgic mood is best captured through sepia coloured photography. This kind of Europeanized café, which gradually became generic throughout the 1950s, is today an object of nostalgia, known as *modan* (the modern of the past), full of grandfather's functional bent metal chairs and padded seats, Formica tables and pressed tin décor – designs sometimes Bauhaus-influenced. Evoking the modern of the 1970s is the café Efish in Kyoto where goods such as modernist tableware and furnishings are sold along with trendy beverages such as blackberry smoothies. Cafés also give nostalgic reference to the Europeanized Taishō period, as exemplified by the Café Sagan, on the eastern side of the Kamo River, where writers and artists, local housewives, local businessmen and office clerks begin the days and end their afternoons. Everyone chats with the master whose collection of French Galle glass lamps sets a vaguely Euro-Japanese tone. Gallery-cafés such as ETW up river in Kyoto, and Café Rihou, on a northern canal, provide a stylish minimalism which shows off the periodic avant-garde exhibitions held there.

Coffee and its places have demonstrated such regional distinctions from the late Taishō period, when Osaka began to be the dominant coffee city in Japan. Coffee houses began to reflect local tastes and the brew strength and quality varied significantly across Japan, as it still does. Osaka is known among coffee people as the 'Three Highs' city: the highest (darkest) roast, the highest density in grams per cup, and the highest rate of consumption. Nagoya uses a medium high roast and 13 grams per cup; Kyoto has a high roast and 14 grams per cup while Tokyo prefers a light roast and 9–10 grams per cup. Tokyo's coffee is said to be almost American, meaning rather weak, as one coffee expert told me.

In all of these places, coffee gained its popular status first as a trend, then as a normal beverage eclipsing ordinary green tea. By the late Taishō era, people considered tea a drink for home, or to end a meal, and coffee the drink for relaxation and entertainment outside the home. *Nihoncha*,

Japanese tea, was like the air (Ishige 2000), something that arrives unbidden when you visit a home or an inn, not something to pay for or sit over.[2]

Tea thus remains a beverage of great significance: as *nihoncha* it is a product with heritage and identity; a healthy drink; and can be a gift of locality from tea-growing areas such as Uji. The majority of tea drinkers also drink coffee, except for some older people who say that they drink only tea for their health. Most coffee drinkers also drink tea at home or at different times in the day. It appears that although in some senses coffee competes with other drinks, overall coffee has simply added an option as well as an activity and a destination, rather than replacing older drinks and spaces in people's lives.

## The Contemporary Scene

Japan has become the world's third largest coffee importing country, and the consumption of coffee leads all social drinks, outselling beer or tea; though there is also an emerging market for non-Japanese specialty teas, while herbal beverages, and *kōcha* (black tea), are ubiquitous in homes. Cafés are abundantly available. There are two or three coffeehouses on almost every Japanese city block and most are well-patronized throughout the day and evening. Most people have a local favourite, local either to work or home. For entertainment, community engagement or private solace, the café is always available. The relatively high price of a cup of coffee in Japan – the range is about ¥250 (about US $2.40, lower in some chains) to ¥2000 (US $19.50) for a fabulous handcrafted cup – pays for the rental of a valuable piece of real estate: the seat in which you talk, read, write, or muse.

2 One exception to the rule that green tea and coffee are non-overlapping is seen in an elegant *kissaten* in Tokyo where a small cup of green tea is brought to the table as the customer peruses the coffee menu, a demonstration of tea's function as a hospitable greeting, even in a specialty coffee shop.

Consequently Japan's annual importation of coffee is about 500,000 tonnes (about 7 million bags). The largest suppliers are Brazil, Colombia and Indonesia for ordinary coffees; among specialty coffees, Hawaii's Kona and Jamaica's Buruman (Blue Mountain) have been consistently at the high end of the market in Japan. In the case of these two coffees, one of Japan's largest import companies has built plantations and commandeered a large proportion of the beans grown in Jamaica and the Kona area of Hawaii to ensure a steady flow for the Japanese market. In the recent past, the demand for these beans has exceeded the available crop yields and there have been reports that cheaper beans from other parts of the Caribbean and Latin America had been sent to Jamaica and Kona, then re-labelled and reshipped as being from those ports.[3]

While Japanese coffee palates are becoming increasingly sophisticated, Starbucks has (as of March 2014) 1,034 locations in Japan, recent reports show that customers find its coffee and presentation too standardized, its service nil, and the cafés too anonymous. Starbucks has begun a restructuring programme in the United States and it appears that growth is slowing in Japan. Currently the largest consumer niche for Starbucks in Japan is young women, who enjoy service at the counter (called in Japan self-service), take-out coffee and what have come to be called coffee desserts – highly sweetened and flavoured coffee-based drinks. In addition, as some reported to me, young women go to Starbucks for the opportunity to meet foreign students and young businessmen who frequent Starbucks.

Older forms of connoisseurship appear to be echoed in specialized niches such as the coffee sipping market, reminding some of the heights of tea-tasting or western wine-tasting. Places such as Café de l'Ambre (Café do Ranburu) in Shinbashi in Tokyo offer rarities for the highly evolved coffee drinker. Owned and managed by Ichiro Sekiguchi (now over 100 years old), Café de l'Ambre is a place where coffee has its own rituals and

3 This could count as fraud. To be designated as a local coffee, at least 30 per cent of the beans must be from the locality. Locality does sell beans – though the bean varietal may be the same as that from a different area. Japanese consumers engage in the *terroir* (specificity of place) cachet of coffee that has engaged Americans as well in a new geographical connoisseurship.

has a sacred aura, though Sekiguchi himself says he hates the tea ceremony because it emphasizes form over flavour. Famous and, to some, notorious for his finickiness, he says his methods are only to ensure the production of the best-tasting cup of coffee in the world. At this specialist café, coffee is roasted twenty-four hours before brewing and ground at the moment of service, while the water reaches a boil. After a short wait for the water to cool – a thermometer will give the server a reading – the server pours the water into a pot with a very fine spout that Sekiguchi has pinched in order to allow only a very thin drizzle, and pours the hot water slowly, in several rounds, over the grounds in a flannel sack. Sekiguchi encourages people to call the day before they visit so that they can request a bean to be roasted for their next day's coffee. He values aged green beans[4] and offers a Yemen '92 as an especially good year. No food is served: he says if you want food, go to a restaurant. Such attention to the coffee, if not to the customer, bears a high price: a typical cup is ¥1500, about fifteen dollars. Such places are not for the young woman who enjoys sipping a Starbucks coffee frappe.

The refinement of taste implied by Sekiguchi's methods might look like fetishism: some say his work is *kodawari* or gone too far. In fact, the concept of *kodawari* is on everyone's lips in the specialty coffee industry in Japan: this is what sets Japanese coffee apart from the rest of the world. High standards, dedication, devotion, a one-hundred-percent engagement with the work of coffee are contained within the notion of *kodawari*. You have to have *kodawari* to be a good producer, coffee taster, and brewer. Taste is the goal, and a café with elegant surroundings, excellent service and good suppliers of beans cannot automatically provide the best coffee to its customers unless it also demonstrates this perfection-seeking dedication.

For those with less stratospheric tastes and thinner wallets there is a local chain, Doutor, which also owns a high end competitor to Starbucks, Excelsior, whose logo mimics that of Starbucks. These chains have provoked both the decline of some local shops and a backlash that has revived interest

4 Coffee can be aged unroasted, but must be drunk within two weeks of roasting, Japanese and western experts insist.

in independent cafés, just as coffee drinking has risen to the higher levels created by coffee industries and cafés.

Japanese drink on average thirteen cups of coffee per week although consumption keeps increasing. This includes coffee drunk in cafés and *kissaten*, in offices, cans bought from *jidohanbaiki* (vending machines, selling hot and cold coffee) and instant coffee, as well as regular coffee drunk at home. Regular means brewed fresh (dripped in a hand-pour, filtered, pressed and other methods) from ground coffee, and this is the most popular form of coffee, though until the 1980s, instant coffee was the leader in Japan.[5] Vending machine canned coffee, and bottled coffee drinks available at convenience stores do not supplant a customer's regular coffee habits. The same person who has just had a regular coffee with a morning set breakfast at a café will take a train to a meeting at another café, and grab a canned coffee on the train platform from a vending machine. The can is not so much 'coffee' as it is refreshment reflexively grabbed en route. The custom of walking down the street drinking a cup of coffee has not taken hold, however. It is rare to see a person with a carry cup, and the older etiquette prevails: you buy a drink from a machine, stay near it while drinking it and place the empty can in a special bin next to the machine.

Espresso and drinks made with espresso such as cappuccino have not caught up to their popularity in the United States. A coffee expert said that it is not artisanal enough, since a machine intervenes between the hand and the brew. One espresso maker, Katsu Tanaka of Tokyo uses the machine, he says, as 'an extension of my hand' and customizes his espresso machine to get the maximum control into his own hand. His clients get to his shop, Bear Pond, early: his method is time and energy intensive and he can only deliver about twenty-some shots of espresso a day. His is a shop for serious aficionados.

5 Neither decaffeinated nor flavoured coffees have found a large audience in Japan. People rarely complain that coffee keeps them awake, and the Ministry of Health has made it very difficult to get an import license for decaffeinated coffee, noting that the processes of decaffeination may produce health hazards greater than any incurred by caffeine itself.

Other cafés are less about the coffee and more about sociality. In places where locals gather, the clientele tends to be made up roughly of age-mates. For example, a six-seat café (two tables, two seats at the bar) tucked into a corner of the old streets of Nishijin in Kyoto is a quiet place. There is a verbal community if its mostly retired, mostly male clientele choose to speak, but silence is also a mode of sharing space. The rising rate of the elderly in Japan (people 65 and older now make up about one-fifth of the population) has made coffee shops very significant gathering places. In some, the regulars check on those who don't show up, in a kind of informal mutual aid.

In university neighbourhoods, cafés become intellectual communities: Shinshindo, a student-faculty café to the north of Kyoto University resonates with lively discussion. However, in the lives of busy people, cafés are not only social spaces; they can be also places to be private in public, to take unmarked time in an unmarked space and for many people this is the primary virtue of the café. As one businessman said to me: 'Taking the time to do nothing – isn't that a wonderful thing?'

Being alone may also mean being unknown to those around you. Learning urbanity also means learning to be in public spaces not identified as 'local' – places where people known to you do not congregate. The new urban spaces give a sense of membership in the global community, an entity with built-in contradictions. Membership here means something fluidly conceived, diversely represented, and offering the freedom to choose new identities. Students, housewives and beleaguered salary workers can imagine the lives of others as they re-imagine themselves in these role-releasing places.

In the fluidity of a café's space, the visitor gains ears, eyes and voice with which to observe, adopt, consume and create new ideas, tastes, and goods, and these styles are not only imported from the 'first world', but are now often imported from other parts of the world, which were formerly considered déclassé, such as Thailand, sub-Saharan Africa and Latin America. Interest in 'ethnic' ways is a result of greater experience in the world, more travel since the boom years of the 1980s, more curiosity and independence for Japanese who travel. The valorisation of the ethnic represents a new globalism, which might be interestingly juxtaposed with political resistance to immigration from these ethnic areas.

It could be said that revivals of older café cultures demonstrate a backlash against the homogenizing forces of Starbucks and other chains. The establishment of new retro-style cafés give reference not to European antecedents, as Starbucks itself tried to do, but to Japanese cafés of the past, to Japanese modernity and Japanese artistic experimentation. There is continuity here with how some of the more outré aspects of the Taishō period cafés were revived in the post-war period but with new international cultural influences. The creation and reinforcement of all kinds of taste in cafés has continued to develop in cafés as performance art of all kinds, musical events, anime presentations and other newer media find venues in coffee shops.

## Conclusion

The consumption of coffee and the construction of cafés in Japan throughout the twentieth century forms an almost perfectly straight upward line of development, save for the war years. By 1930 there were 800 cafés in Osaka alone, and by 1933 there were 37,000 cafés in Japan and now roughly number over 80,000 nation-wide. Currently, the specialty coffee market in Japan is burgeoning, here specialty does not mean coffee desserts such as the Starbucksian caramel macchiato drink, but coffee whose provenance and care is of great worth and note – the best of a crop. Japanese speciality coffee standards are known to be very high, so that another form of globalization has ensued: if a new varietal appears, or a new *fazenda* (coffee plantation) wants to put its beans on the international market, samples of beans are vetted first by Japanese consultants. If it can sell in Japan it can sell anywhere, at least that is what the coffee specialists of America and Europe aver.

Japan's effect on the global coffee industry is therefore profound, from the creation of a Japan-Brazil nexus creating the first large export market in Japan for what is now the world's leading producer, to the engagement in coffee development projects worldwide. The very high standards of coffee drinking in Japan have spurred coffee producers to meet the challenge of the Japanese buyer – the specialty coffee producers say they are working for greater uniformity in the beans in quality and sizing, better fermentation and drying processing, better storage and shipment methods – all because of Japanese buyers.

Consumption in Japan is not only about the quality of the beverage itself, it is about the spaces where it is consumed. Cafés serve important functions in Japan, some particular to Japanese society where locations such as school and workplace can demand much from an individual, in roles and in rigid performance standards. In addition cafés have been the crossroads for the global flows of culture and politics; globally sourced coffee and foreign cultures and behaviours have defined such places and the people who frequented them. In the early twentieth century, the frequenters of cafés took on the cachet of cosmopolitanism, but now, in the early twenty-first century, a coffee aficionado is a connoisseur of a commodity both global and domestic. The consumers of worldly brews, as frequenters of cafés find themselves (in several senses) in local spaces which epitomize comfort and sociality in very Japanese terms.

## References

Barthes, R. (1987). *Michelet*. Trans. by R. Howard. Berkeley: University of California Press.

Cwiertka, K. (2003). 'Eating the World: Restaurant Culture in Early 20th Century Japan'. *European Journal of East Asian Studies*, 2(1): 89–116.

Dalby, L. (1983). *Geisha*. Berkeley: University of California Press.

Elias, N. E. (2000). *The Civilising Process*. Oxford: Blackwell.

Ellis, M. (2004). *The Coffee-House: A Cultural History*. London: Orion.

Habermas, J. (1991). *The Structural Transformation of the Public Sphere*. Cambridge, MA: MIT Press.

Hoshida, H. (1988). *Nihon Saisho no Kōhiiten*. Inaho Shobō.

Ishige, N. (2000). *The History and Culture of Japanese Food*. London: Kegan Paul.

Mori Company (2002). *Mori Building Vision*. Mori Company.
Naomichi, I. (2001). *The History and Culture of Japanese Food*. London: Kegan Paul.
Silverberg, M. (1998). 'The Invention of the Modern Café Waitress'. In S. Vlastos (ed.), *Mirror of Modernity: Japanese Inventions of Tradition*, pp. 208–28. Berkeley, CA: University of California Press.
Terada, T. (1933). 'Kōhii Tetsugaku Jōsetsu'. Keizai Ōrai.
White, M. (2013). *Coffee Life in Japan*. Berkeley: University of California Press.
Yokomitsu, R. (2001). *Shanghai*. Ann Arbor: University of Michigan Center for Japanese Studies.

CHRISTINE R. YANO

# 'A Japanese in Every Jet': Globalism and Gendered Service in the Jet Age

> I always dreamed:
> What would it be like
> on the other side of the horizon?
> —HIROKO (PAA 1966–1972)

## In the Beginning

On 1 April 1964, the Japanese government lifted the international travel restrictions it had imposed since the days of the American Occupation (1945–52), opening the travel floodgates for Japanese citizens. With rising incomes and the broadened aspiration to participate more fully as global citizens, Japanese tourists began arriving on foreign shores from Honolulu to London. The number of Japanese travelling abroad increased yearly, beginning with 128,000 in 1964 and expanding to more than five million by the late 1980s. This included not only tourists but, more commonly, businessmen.[1]

To accommodate this travellers' market, international air carriers began hiring Japanese women as stewardesses.[2] This practice worked as a

1 Some of these men were travelling to South America to establish and maintain business connections through existing Japanese emigrant networks.

2 I use the term stewardess, primarily because this label more accurately reflects the historical period. Moreover, some stewardesses took pains to correct me, indicating that they were stewardesses at the time, not flight attendants.

marketing device in multiple ways: not only could Japanese stewardesses provide a comforting in-flight presence for Japanese travellers, but for everyone they added cosmopolitan glamour and proof that these airlines truly represented the world. The 1960s was not the first time that a foreign carrier had hired Japanese stewardesses; in the 1950s various airlines had experimented with hiring Japanese speaking women, including Japanese Americans, in order to compete with the newly established Japan Airlines.[3]

The 1960s – with increasing Japanese passengers, flights, and an overall rise in foreign travel – was the first period during which airlines hired Japanese women on such a large scale. As an indication of this practice's magnitude, the 1 May 1967 issue of *Life* magazine featured a cover story proclaiming: 'Newest Stewardess Fad: A Japanese in Every Jet', with a multi-page photo spread of Japanese stewardesses from eleven international carriers.

Standing dead centre in both the cover shot and inside photo montage was Hiromi Abe, a Japanese stewardess with Pan American World Airways (from here on: Pan Am) her placement in the photos was no accident. At the time, Pan Am was the undisputed leader, trendsetter and global aviation's prestige symbol. Here I examine Abe's and other Japanese stewardesses' experiences flying for Pan Am as part of the phenomenon that *Life* magazine dubbed 'A Japanese in Every Jet'. This phenomenon provides a case study of globalism, gender, and modernity during Japan's initial period of cosmopolitanism which coincided with much of that of the industrialized world.[4]

3 The history of Japanese stewardesses on foreign carriers begins in 1951 when Thai Airways hired Japanese women. Their employ only lasted one month due to lack of Japanese customers. In 1952 Northwest Orient Airlines hired Japanese stewardesses for its Tokyo-Pusan flights, and continued their employ in the following years. In 1955 three other foreign carriers hired Japanese stewardesses: Air India International, KLM Royal Dutch Airlines, and Air France (*Nippon Times* 6/29/55; 9/20/55). A photo in the *Nippon Times* (September 20, 1955) depicts seven Japanese women flying internationally as stewardesses on the following airlines: Pan Am, Northwest Airlines, KLM, Japan Airlines, Air France, CAT, and Air-India International.

4 I have discussed these issues with somewhat different emphases elsewhere, specifically in Yano 2013 in which I emphasize mobility, modernity, and gendered labour,

Here the term cosmopolitanism is used to refer to an era, an airline, and persons. Hannerz discusses cosmopolitanism as an orientation, 'a willingness to engage with the Other', and a matter of competence: 'a personal ability to make one's way into other cultures' (1990: 239). According to Hannerz, travel alone does not make one a cosmopolitan; rather, one must be willing to engage on a level that goes beyond shopping and beaches. He thus creates a tripartite formation of locals, cosmopolitans, and tourists (1990: 242). This case study embeds Hannerz's approach to cosmopolitanism within its historical arc.

Pan Am embodied cosmopolitanism not only through the distances it travelled, but also through the confidence it established in its experience of the world. Its slogan – 'World's Most Experienced Airline' – pinpointed this exactly. As the creator and exemplar of the Jet Age, Pan Am was the airline of choice of that cosmopolitan new breed: the jetsetters. This included world leaders, movie stars, and other celebrities, whose use of Pan Am enhanced their images, as well as the airline. In fact, the airline itself acted as a celebrity, dominating the world's skies from the 1940s through the 1970s. Thus I examine the intertwined themes of celebrity and cosmopolitanism, at both the corporate and personal levels, asking:

1) What corporate strategies were involved in the hiring of Japanese women?
2) How did Pan Am's Japanese stewardesses view international travel and employment by Pan Am?
3) How were gender, race, class, and cosmopolitanism inflected in these meanings and practices within the context of the post-war era?

Through interviews and archival research, I analyse these women's role as early pioneers of globalism in Japan. When conducting interviews with people recalling their experiences from decades past, we must take their

discussing stewardesses as forms of 'modern girls', and thus contributing to that volume's title *Modern Girls on the Go*. However this chapter gives a broader overview of the phenomenon of Japanese stewardesses hired by Pan Am.

recollections as configured from the perspective of present-day lives. This is neither a limitation nor advantage, but simply part of the context of their memories. They often spoke of their gendered service as racialized (sometimes dubbed as 'flying geisha'), requiring a tightrope performance straddling tradition and modernity. Notably, the women used their position for their own purposes, gaining a foothold in cosmopolitanism in ways seldom shared by Japanese men of the same period.

## Global Citizenship, Gender, and Social Class

The hiring of Japanese women by Pan Am and other international airlines occurred within the context of Japan's dramatic entry into global citizenship.[5] Flying Pan Am signalled Japan's modernity by way of the Jet Age – an era enabled by growing middle-class prosperity, defined by mass tourism and the celebration of technology. Gender played no small role in this aspect of global citizenship. A 1967 *Life* magazine article provided a litany of qualities that purportedly made Japanese women particularly suited to the stewardess job:

> While the Japanese airlines have always been aware of the attributes of Japanese women, the foreign carriers only lately have discovered their unique talents. 'It is her air of serenity and gift of grace that makes the Japanese stewardess such a sought-after member of the airlines' crew,' says one Lufthansa official. An Air India representative adds admiringly: 'Their ability to stand up to strain better than others is a major asset.' … 'A Japanese woman knows how to serve and desires to serve,' says Ursula Tautz, Lufthansa's chief of stewardesses.'
>
> (*Life* 1967: 42–3)

5 The other airlines listed in the *Life* article included Qantas, BOAC, Scandinavian Air Service, KLM, Air France, Lufthansa, Air India, Northwest, Alitalia, and Cathay Pacific.

1960s Japanese stewardesses flew as exemplars of tradition, even when – or more to the point, especially when – engaged in that very modern of professions, the international flight attendant.

Part of this gendered juxtaposition of tradition and modernity resides in the nature of the stewardess' job, which calls upon servers to reassure passengers about the comforts and safety of flying, especially in the 1960s when jets were new. A large part of the flight attendant's work rests in the management of passengers' sense of wellbeing (Hochschild 1983), affirming the skies as a safely domesticated space. As the historian Barry explains: 'A stewardess's foremost duty was to mobilize her nurturing instincts and domestic skills to serve passengers, much as middle-class, white women were expected to treat guests in their own homes' (2007: 12). If the airplane cabin was supposed to be one's living room, then the flight attendant's job was to emphasize its homey-ness, rather than the technological superiority of the jet.

These elements of the flight attendant's job during the Jet Age were further refracted when Japanese travelled abroad. For Japanese the Pan Am cabin could never be quite their own living room; rather, it represented a borrowed space of white, American-based privilege, modernity, and globalism. Having Japanese stewardesses on board brokered the gap between Japanese passengers and that borrowed space. For Japanese and other passengers alike, the link between the Japanese stewardess and the conservatism of 'tradition' acted as significant, gendered selling points. The best way to be modern for the Japanese stewardess lay in performing the traditional roles of service and hospitality, even while high in the air aboard the latest in technology. For the Euro-American traveller, she was the exotic, jet-age hostess; for the Japanese traveller, she was the home-away-from-home domestic ideal.

These issues of gender also intersect with social class. Passengers flying internationally during the 1960s were typically upper or upper middle class, whose expectations of service were high. The Japanese stewardesses whom I interviewed emphasized the class position of the (primarily non-Japanese) passengers, evidenced by the way in which they dressed and their generally well-mannered behaviour. The fact that class relations paralleled hierarchical relations between the Japan and the United States only amplified the status gap between the Japanese stewardess and her American employer as well as between other American stewardesses and passengers. Furthermore, more men than women travelled, often for business; therefore expectations of service followed intertwined gender and class lines. The demands placed upon Japanese stewardesses came not only from the company and its training, but also from passengers' structural position which was based on class, race, nation, and gender.

The Pan Am cabin, in effect, served as a newly created space of interaction built upon old practices and assumptions. Pratt's (1992: 24) concept of contact zone is useful here, defined as 'social spaces where disparate cultures meet, clash, and grapple with each other, often in highly asymmetrical relations of domination and subordination.' The disparate cultures of Pan Am cabin's contact zones included: Japanese stewardesses and the US/Pan Am corporate culture; a polyglot of upper class passengers from various nations, including Japan and the United States; Japanese, American, and European stewardesses. The 'meeting, clashing, and grappling with each other' in this contact zone occurred within ongoing US-Japan relations, beginning from the Second World War, to the American Occupation and through Japan's economic recovery. The 1960s global cabin as contact zone performed its modernity as an assemblage, spinning the asymmetries of post-war relations within the frame of a higher order of corporate culture known as Pan Am.

## Cosmopolitan Dreams: An Appetite for the Foreign

Although Pan Am had been hiring Japanese American stewardesses since 1955 for their Japanese language skills, with the easing of US immigration laws in 1965, Pan Am could tap into a wholly different source by hiring Japanese nationals. The first seven Japanese women hired by Pan Am in 1966 garnered considerable media attention. A veteran Pan Am pilot who ended up marrying one of the women recalled: 'Pan Am got a tremendous publicity boost in Japan; those seven girls' pictures appeared in papers all over the country.' The publicity worked in multiple ways, as promotion not only for the airline, but also as evidence of Japanese women's prominence on the global stage.

Who were these women and what kinds of meanings, dreams, and desires did they give to Pan Am employment? Examining this focuses the discussion on the lived experiences of globalism. As Appadurai argues: 'Fantasy is now a social practice: it enters in a host of ways, into the fabrication of social lives for many people in many societies' (1996:198). The fantasies of the women I interviewed, as well as the narratives compiled by the Tokyo branch of World Wings, International (an organization of former Pan Am flight attendants), focus on freedom, of wanting to go beyond Japan in order to know and experience what was on 'the other side of the horizon', as quoted at this chapter's beginning. I situate these dreams within generational, class and national positions – that is, growing up in post-war Japan in a middle or upper middle class family. While these women shared some of these dreams with many Japanese in this period, their social class placed them in a better position to realize them. Indeed, class privilege conferred greater access to places, cultures, and practices that both predated and enabled their Pan Am employment. Akemi (PAA 1968–86),[6] for example, describes her family background: a geologist father who travelled worldwide and lived in Singapore while she, her mother and siblings lived in Tokyo; as well as an aunt who attended university in Los Angeles and married a non-Japanese. For a family like Akemi's, travelling to and even living in foreign countries was not an anomaly, but an expectation.

Cosmopolitanism in these upper class families often ignored gender. Although privilege was not gender neutral – with sons of families typically receiving higher levels of education at four-year universities, compared with daughters – travel opportunities for women allowed them to break societal

6 All names are pseudonyms, except for those whom I identify in conjunction with their specific historic contribution. I also provide dates of Pan Am employment parenthetically, for example (PAA1968–1986). Many women's dates end in 1986, when Pan Am relinquished its Pacific routes to United Airlines.

barriers, violate conservative gender codes, and leave Japan by themselves. Idiosyncrasy may inhabit this class territory as well. When I met Sumie (PAA 1967–72) in May 2005, I was struck by her appearance: short and wiry, she wore not a trace of make-up, had her hair pulled into a pragmatic, greying ponytail, and described her hippie lifestyle living in rural O'ahu with chickens and goats. Through conversation I came to interpret her unconventional lifestyle as enabled by the confidence and expectations borne of class privilege. She recounted her family's background of generations of cosmopolitanism:

> In my lifetime I always knew that I would be coming to America because my grandparents, my grandfather had been here. So had my grandfather's younger brother, and my grandmother's younger brother was also in England. In my family people have gone in a hundred years before to America, to England, to Germany for various study and work.

Sumie's background not only raised expectations for international travel, but also provided the support and role models for her to act upon those dreams. She says that her family's reaction to her Pan Am employment was extremely positive: 'They were excited, very. They backed me up 100 per cent.'

While some families, like Sumie's, reacted enthusiastically to their daughters' international aspirations, others did not. Kumiko Sato from the first group of stewardesses in 1966 recalled:

> I told my family that I wanted to become a flight attendant and they strongly opposed this. I was not expecting such a reaction from them. Their reason was, 'It is manual labour. Serving meals and drinks for a living is out of the question!' I had never experienced such rejection by my family before, so I was in a state of deep shock. … After much deliberation, I decided to seek advice from my uncle on my mother's side, and thus I began my lonely battle with my family. Thanks to my uncle's help, my persistence was rewarded with my family agreeing to let me fly for one year. (2004: 102)

The stewardess job occupied an ambiguous position, sharing practices of travel overseas with the elite, but locked into the lower status of a server.

Japanese flight attendants were educated; some had even lived overseas as exchange students in the United States. They all could speak a level of English that was unusually high for the time, and demonstrated a familiarity with Euro-American culture that was in and of itself class-based. Some had majored in English in college; others had attended English-language schools after graduation. In choosing to fly for Pan Am, these women were opting out the expectations of marriage and motherhood at home, in favour of foreign employment. Many negotiated this uncertain terrain by assuring their parents that their employment would be temporary, usually for one or two years. However for others, this became a far longer break. These include the Japanese stewardesses I interviewed, now living in Honolulu where they had been based at Pan Am's Asian-language headquarters.

On an individual level, choosing to be a flight attendant with Pan Am suggests certain personal qualities: an appetite for the unknown and a willingness to forego, if temporarily, a housewife's or secretary's lifestyle. These structural and personal attributes imply either running toward 'the other side of the horizon', as well as running from or rejecting what was on 'this side of the horizon': post-war Japanese society. Ironically, this rejection entailed performing at least some of the duties of the stereotypical Japanese woman.

Personal vignettes illustrate the ways in which women narrate their negotiations with the demands of Japanese society. Hiroko (PAA 1966–86) from Kita-Kyushu described herself as a 'vagabond with wanderlust', fascinated with foreign cultures since childhood. She recounted: 'I daydreamed all the time whenever I saw planes taking off. I was intent upon travelling the world. I always dreamed: what would it be like on the other side of the horizon?' This hunger for the 'other side of the horizon' included Italy and Hawai'i. She learned about Italy through her older sister, who had studied European art. Thumbing through her sister's art history books, she was particularly fascinated with Michelangelo's paintings, and vowed to visit Italy in order to see his work. While at Doshisha University in the 1960s majoring in English literature, she also experienced the 'Hawai'i fever' that engulfed Japan at the time, especially with films such as Elvis Presley's *Blue Hawaii* (1961), and popular Japanese actor Yuzo Kayama's *Hawaii no Waka Daisho* (Hawai'i's Young Guy) (1963) and older songs such as '*Akogare no Hawaii Koro*' (Dreams of Hawai'i) (1948). Hiroko herself learned to play Hawaiian songs on the ukulele. She said: 'Everyone's dream was to go to Hawai'i!' In pursuit of her goal, she took a job as an airport hostess at Haneda International Airport. There, she watched the stewardesses come and go, observing the ways in which they dressed and acted. This then led to her first employment as an international stewardess with Qantas. Hiroko says that it was with Qantas that she first learned how to perform as a particular kind of Japanese woman – a geisha-style hostess to foreigners. When a job with Pan Am became available, Hiroko jumped at the chance. For her, Pan Am represented an opportunity to fulfil two of her lifelong dreams: visiting Italy and living in Hawai'i.

Another former Japanese stewardess entering Pan Am employment in 1966, Noriko, credits curiosity with drawing her to Pan Am. She said: 'I was always interested in anything going on outside of Japan. My curiosity level was far beyond average. Plus I enjoyed staying in Michigan as an exchange student back in 1962 and thought it is the best way to travel if I became a flight attendant.' Noriko's desire to see things 'outside of Japan' caused her to reject considering any job with domestic airlines. She explained: 'I always wanted more freedom than [what] the Japanese carriers [would allow me].' Another Japanese stewardess explained: 'I wanted to travel to foreign countries right away, so I decided to join Pan Am which had the most flights overseas at the time' (Suzuki 2004:110).

For these women, Pan Am represented freedom through cosmopolitanism. Sumie noted: 'I looked at Pan Am as very luxurious. Everything about Pan Am was bigger, fancier, more history.' By this, she means that as the dominant international American carrier of the period, Pan Am played a central role in historic events, such as airlifting military personnel and orphans out of Vietnam. The comparison she makes – 'bigger, fancier, more history' – references Pan Am's status and prestige during this period. Flying with Pan Am meant absorbing those qualities as personal attributes. Several stewardesses mention that Pan Am was symbolic of the USA itself. One woman who flew in the 1970s reminisced: 'I believe Pan Am was loved all over the world and was symbolic of the United States!' (Arima 2004:17). Working for Pan Am meant moving to the USA and gaining a certain form of American citizenship – not in political terms, but in terms of membership in and identification with an American corporation. As Saeko (PAA 1971–81) pointed out: 'You have to remember, the 1960s, 1970s, America was up there. Rock and roll, Elvis Presley.'

Indeed, the airline's image went beyond America. Hiroko gushed: 'Pan Am was like a flying United Nations!' In this, she emphasizes the following: Pan Am flew the most extensive international routes of all carriers; the airline's passengers included people from many countries; and its crew was comprised of not only Americans, but also Europeans and Asians. Her characterization, in fact, belies more than these literal aspects of the internationalism; instead, it points to American domination of the international sphere – astutely paralleled by her reference to the United Nations. As she spoke, she referred to Pan Am's globe logo. 'That is the Pan Am world, the earth, and as a stewardess, I am hostess to that globe.' Hiroko accepted and even lauded Pan Am's corporate ambitions as her own.

Connecting Pan Am with Japan and the world came as no accident. Besides its presence in the skies, Pan Am also created its own media spectacles in Japan as elsewhere. One of these was the presentation of a major trophy at the nationally televised sumo tournaments yearly from 1961 to 1991. The sumo-watching public came to know Pan Am through its Far East Regional Manager, David Jones, a white man who presented the Pan American Trophy annually.[7]

However, more important in the process of linking Pan Am and the world was a television show broadcast on Tokyo Broadcasting System Network (TBS) from 1959 to 1990, *Kanetaka Kaoru Sekai no Tabi* (Kaoru Kanetaka's The World Around Us Travelogue), with group support from Pan Am. In this programme, an attractive, English-speaking, Japanese woman, Kaoru Kanetaka, featured different exotic travel destinations – flying Pan Am. Throughout its thirty-one years' run, Kanetaka visited more than 150 countries on all the continents, from the deserts of Africa to the

7 Jones had acquired such celebrity in Japan specifically as the Pan Am sumo presenter that his death in 2005 was noted in many Japanese newspapers; see his obituary in the *Japan Times* (02/06/05).

ice-covered North Pole (Nakata 2001). Akemi (PAA 1968–86) recalls this programme's impact and the way in which it fuelled her Pan Am dreams:

> *Kanetaka Kaoru Sekai no Tabi!* Everybody watched it. She was a very beautiful lady, and very elegant, and she goes to different countries, she interviews people, she goes to the city, she goes to shopping areas, she goes to different resort areas, and she just introduced all different countries, and it was always with Pan Am. Of course the Pan Am logo comes up, because Pan Am did that. It was Pan Am that made the program I used to watch. And everybody just dreamed about Pan Am. That was the dream airline! That's the number one airline everybody wanted to work (for) or fly (on).

Pan Am in Japan thus produced the media visions of the world that made it the 'dream airline'.

The dream of flying for Pan Am was not without its economic merits. As Saeko (PAA 1971–81) pointed out, when she was hired:

> The yen was 360 to the dollar. So my first salary, including per diem and all that, was close to a thousand dollars a month. But when I calculated it in Japanese yen, it was a lot of money! About 360,000 yen. And my classmate, his starting job was about one-fifth of that!

These lifestyle benefits combined with the comparatively good pay made the job highly desirable for Saeko and others.

## Flying the Pan Am Skies as a Bodily Performance of Race and Class

For many Japanese stewardesses the initial Pan Am encounter began as a meeting of bodies in training – and therefore of matters of race. To begin with height was an issue: Pan Am had lowered its height minimum from five feet two inches (1.57 meters) to five feet one inch (1.55 meters) especially for the Japanese recruits, and many of the women barely reached that. In Japan, the stewardess profession was thought of as a job for tall women, partly because of the need to reach overhead bins and other features of planes manufactured in the United States, partly because of modern images of beauty. In fact, Japan Airlines had more stringent height requirements than Pan Am. Sumiko, for example, had been rejected by Japan Airlines because of her height of just under 5'1", before being accepted by Pan Am. However it is not so much that the rules of the two airlines were different; it was more that Pan Am was willing to bend the rules further in order to hire Japanese stewardesses.

Being shorter than many of the other stewardesses affected the ways in which the Japanese stewardesses thought about themselves: placed in a setting of taller (white) women, it was difficult not to think of being short as a shortcoming. Saeko (PAA 1971–81), for example, recalls landing in Miami for training and being stunned by all the instructors, tall European and American recruits: 'All the instructors were very pretty, slender, tall. It's the typical Pan Am image, and I go – omigod!' Amongst the taller women (and people, generally), she felt intimidated. It is not as if all Japanese stewardesses were short, nor were all non-Japanese stewardesses tall. However many Japanese stewardesses apologetically mentioned their short stature to me as a significant part of their initial encounter with Pan Am, which suggests its symbolic importance.

In training and on the job, Japanese stewardesses could compare their bodies with that of other (white) women. They were assisted in this by some European and, to a lesser extent, American stewardesses' practice of shedding their clothes when in the hotel rooms where they stayed during layovers. Several Japanese stewardesses recalled their shock at this uninhibited practice:

> You share a room with a Caucasian girl who would just basically take off all her clothes in the room! I mean, in Japan, even with a roommate, you just don't take off your clothes. You cover yourself. But some of the European girls were just hanging around naked! And I was just like – omigod! I have to share a room with these girls? But after a while you get used to it.
>
> (Saeko)

Often what the Japanese stewardesses perceived only confirmed some of the racial stereotypes by which Japanese viewed their bodies as inferior – that is, shorter, thinner, weaker (and smaller-busted) – to those of whites (cf. Frühstück 2003: 18–19). This comparison formed part of the *akogare* (longing, yearning, desire) for the West, and more specifically for America – defined as a land of whiteness, power and long legs. Kurotani (2005: 167) points out: '*Amerika* is not just a Japanized pronunciation of the English word. It is a complex historical formation of the cultural and political other, a mirror against which Japan came to imagine its own totality as a modern nation-state.' The imagining of *Amerika* was not neutral, but laden with *akogare*. More than one Japanese stewardess remembered being impressed by American television shows, such as *Father Knows Best* and the *Donna Reed Show*. Saeko recalled:

> I remember watching *Donna Reed Show*. She shows up using a vacuum cleaner, with a skirt like this [demonstrates a full skirt], and wearing high heel shoes, and impeccable hair, and I thought, my god, is that what they do? I was curious! ... It's definitely *akogare* [longing], but I didn't want to be Donna Reed. I wanted to see her and see if they really do that.

Training and flying with Pan Am meant learning the accoutrements and practices of the Donna Reed lifestyle. Pan Am provided these Japanese women with the opportunity to put themselves in the frame, and thus inhabit the life of their dreams, both American and global.

More prosaically, those dreams included serving and eating foreign food and drink. Sumie recalls: 'The cooking training was most interesting.[8] They would have local people [in Miami] come in to eat what we cooked. So it was for real, we had to make the food as we would on the plane.' I asked if she had ever previously eaten the foods she was being asked to prepare and serve on board, to which she replied, 'No!' The foreign meals

8 The amount of food preparation required of stewardesses has differed greatly over time. During the epoch under discussion stewardesses had to do a limited amount of on-board cooking, such as scrambling eggs, and a greater amount of re-heating and plating.

included items such as Cornish game hen, beef Wellington, caviar and a variety of cheeses.

That cosmopolitanism included not only the food, but also how to serve it and, during off-duty hours, eat it. Saeko recalls:

> I didn't know how the American people eat. We had some forks and knives at home, but we hardly used it. We hadn't eaten steaks. So Pan Am training school was just totally a finishing school for me to go into this – this America.
> CY: Can you give some examples?
> SB: When you work in the galley, you were taught, cold things were served cold, hot things were served hot, so you heated up the plate and tray. But the salad plate and salad forks were chilled.
> CY: And you'd never heard of that before in Japan?
> SB: No, of course not!

Saeko's narrative illustrates a gap between even the upper class families of Japan and the Pan Am, Euro-American, upper-class culture. Families in Japan might have purchased forks and knives, but they were rarely used. Neither had they eaten much of that quintessentially American food – steak.

Her characterization of Pan Am as a finishing school is a sentiment echoed not only by Japanese stewardesses, but many other women employed by the airline. Few of them had lived a life of such upper-class refinement. Besides cooking, Pan Am taught them how to dress, do their make-up, hair and even how to converse. However this finishing school carried particular resonance for the Japanese stewardesses: they were learning the ways of power where race (white), nation (America) and class (upper) meshed. Thus, learning about chilled salad forks meant adhering to a set of performative expectations that permitted proximity to those in global power. Gender works critically here as well: finishing schools polish women more than men, because women provide the decorative backdrop for power settings. Rather than sitting at tables of power, women in the 1960s and 1970s more typically set them and served the dishes to be eaten there. Pan Am trained its stewardesses to set and serve that table properly as part of the performance of prestige of the airline.

The Pan Am finishing school taught the ways of power that included upper-class, European practices of worldliness. This included not only food, but also drink. Saeko recalls:

> I came from just a decent middle class, or maybe upper middle class family in Japan. So I never drank wine at that point. In the 1970s nobody [in Japan] drank wine. They all drank beer and Scotch. So I didn't know how. I didn't know so much – the way you drink is, you have the cocktail leisurely. … And after dessert, there's a liqueur cart. You know, Crème de Menthe, and Cointreau, and Drambuie. So all these liqueurs! And people will ask, what year? What year is this Chardonnay? And I'd go: 'Huh?'

She was not alone. Many women – Japanese and non-Japanese – expressed bewilderment and sometimes frustration at the barrage of new things they had to learn while in training. However, that bewilderment and frustration intermingled with excitement at this proximity to living their cosmopolitan dreams. For Japanese stewardesses, every time they walked into a house overseas or even a hotel room without taking their shoes off – against Japanese strict rules of purity and cleanliness – they both cringed and thrilled at the violation. This is exactly how Donna Reed lived, walking on a carpet indoors in high-heeled shoes, wearing a shirtwaist dress, hair neatly coiffed, serving dinner on a white tablecloth with silver in place. Both thrill and violation defined the cosmopolitan experience. This was outside their hometown Japanese boundary (violation) in new territory that carried global cachet (thrill).

As stewardesses with Pan Am, these women were able to 'borrow' the company's prestige (Wright Mills 1951) – or, as I term it, they gained the company's 'proximal prestige' (Yano 2011), as well as of those with and for whom they flew. That is, they adopted the prestige of the company they kept – Pan Am, its workers, its customers – by proximity and training. That prestige inhered within class and cosmopolitanism, if not race.

As they adopted the company, so, too, the company adopted them. Immaculately groomed in their blue serge uniforms, they became part of the Pan Am global spectacle. For example, Saeko recalled:

> I didn't quite realize how prestigious Pan Am was until I went to training school. One of the things they [Pan Am school management] call and ask [sic] me is, after

> they take all these graduation pictures, which local [in Japan] newspaper would you like us to send this newsworthy thing? I mean, of *me* becoming a Pan Am stewardess!

Pan Am recognized the value of advertising through the photographs of hometown girls now in their employ, and did not neglect such a public relations moment.

Ironically, they became representatives of not only Pan Am, but also the United States. One woman recalls:

> The day of the moon landing (21 July 1969, Japan time) .... the Pan Am branch office called me on an urgent matter ... They wanted me along with others to be part of the television program to watch the news from the moon. The Pan Am crew appeared in uniform on the program. I thought I was really lucky to be there to witness the astronauts on this historical moment ... It simply made me cry with joy!
>
> (Kawauchi 2004:74)

Her narrative highlights ways in which Pan Am stewardesses, including Japanese, became media representations. More precisely, they became the mediated presence of Japanese achievement: wearing a prestigious American company's uniform, taking on a profession at the forefront of globalism, occupying a position in close proximity to the highest achievements in American technology. This was a public relations moment for Pan Am, Japan, and Pan Am in Japan. The tears she shed represent not only that moment and her place in history, but also her entire experience of cosmopolitanism by way of Pan Am as a physical, mental, and emotional rush.

## Pan Am Employment as Commentary and Critique in Post-war Japan

Many women whom I interviewed spoke glowingly of Pan Am as the major influence that changed their lives. Hiroko declared: 'This is our lifetime love affair! Pan Am treated us so well!' The company *did* treat them well:

paying them generously, providing a per diem allowance during layovers, and housing them in luxurious Pan Am-owned Intercontinental Hotels. A number of women recognize that Pan Am's 'treating employees well' led to the company's downfall in 1991. Equally important for the women was the exposure to different cultures, peoples and ways of thinking that made the experience life-changing. Saeko recalled:

> The European stewardesses, they had their own opinions. I was impressed. They spoke English with their own opinions. ... Me, I didn't have any opinion. I didn't know. I was so naive, I didn't have any opinion. I didn't know the worldly things. That's basically what it was. I had such a sheltered upbringing.

She contrasts Japanese attitudes tied both to gender (i.e. females should be 'sheltered'; they should not have opinions, or at least not express them publicly) and culture (i.e. individuals should not express opinions strongly or openly) with European attitudes that encouraged the cultivation of individual opinions and their expression. Given these juxtaposed attitudes, she credits Pan Am with encouraging her to cultivate herself as an individual – that is, one who could more readily live and express herself on a cosmopolitan stage.

Hiroko calls Pan Am her teacher in global humanism: 'Pan Am taught me that people [around the world] are [basically] the same. No money could have bought us that education! They taught that the world is one family. With Pan Am, I am an earth being.' But to what extent is she a particularly *Japanese* 'earth being?'

When I met Hiroko, she graciously treated me to an elegant Japanese lunch in a secluded restaurant in Honolulu. She brought up the subject of geisha and the ways in which some media likened flight attendants and, in particular, Japanese stewardesses, to 'flying geisha girls'. According to Hiroko, being called a geisha might be taken as a compliment, because a true geisha is one practiced in the arts of service. The ultimate in service is being able to anticipate a guest's needs. She said: 'The essence of service is that you don't have to wait for anything.' However, Hiroko also pointed out that being a geisha in Japan signifies that you are a person from the lower classes, because poor families sold their daughters to a geisha household.

Hiroko concluded that calling stewardesses geisha girls could be taken as both the highest compliment, as well as a low snub. In sum, she sidestepped critiquing both Japanese society with its geisha system and the Euro-American media that labels stewardesses geisha.

Generally, the women I interviewed similarly tended to avoid critique. There are several ways to interpret this. This was a self-selecting group, willing to be interviewed for this research or choosing to write their memories of Pan Am for a retired employees' organization. In the interview context, they may have seen their primary function as that of imparting gracious comments to me. According to several women, stewardesses for Pan Am were generally very happy with their job. They wanted to be there with a prestigious company in a prestigious occupation. Even while training, as Akemi put it: 'I don't think anybody talked back. Everybody was very obedient because everybody wanted to be flight attendant. They wanted to graduate, so I don't think anybody objected to anything. They followed all the rules.' She went on to explain: 'Japanese people are taught to follow the order; since we were little, follow the parents' order.' In this way, not only were most recruits obedient, but Japanese recruits were perhaps even more obedient than most. A lack of criticism may also be in part a pattern of avoiding negative comments when speaking to a relative stranger such as me. Usually pleasantries and celebratory stories guided our conversations.

Although they were generally not critical of Pan Am, I believe that their actions suggest an embedded critique of post-war Japan. These actions took them from Japan at a time in their lives when many of them should have been getting married and having babies – following prescribed patterns of Japanese adulthood. As Massey points out: 'Many women have had to *leave* home precisely in order to forge their own version of their identities' (1994: 11; italics in original). Instead of marrying and having babies, they became working women, employed by a foreign company, living in a foreign land, and inhabiting the liminal world of foreign travel for a living. In parallel with the internationalist women who favour liaisons with white men studied by Kelsky in the 1990s, the potential for transgression and transformation always lies in waiting (2001). If these stewardesses were rejecting post-war Japanese society, they were doing so in the most fundamental way possible – with their bodies. These actions held particular sway because these are the people in whom societies often place some of their greatest investment – unmarried, middle- and upper-middle class women.

Granted, many of these women took positions with Pan Am for only a few years, before marrying and having children. However most of those whom I interviewed stayed in the United States and ended up marrying white men. Their total rejection of the socially prescribed path makes them rebels by default, even when graciously serving green tea and even when acceding to their families' dreams. Other former Pan Am stewardesses whom I met in Japan had chosen a more conservative path, often marrying Japanese men. Their negotiated subjectivities – tacking between their families' expectations and their own worldly experiences – created a differently refracted cosmopolitanism. Still avidly speaking English, relishing their Pan Am memories at yearly gatherings in Tokyo, and maintaining connections through reunions and correspondence with long-held Pan Am friendships around the world, they bring this level of outward-looking orientation to their Japanese upper-middle-class (or upper-class) lives.

I asked Sumiko if her family objected to her marriage to a white man:

> No, no, no. In fact my father ... would say that being short was a handicap. ... So for me to have children who would be taller, would be to marry outside of your race. That's what he said. So ... to make our race better, he said, we have to mix ... So when it came time to tell him that I was marrying my husband, a white person, 6'6", I knew he wouldn't object.

Sumiko's father's position may have been extreme, but perhaps not so extreme as one might expect, given Japan's modern emphasis on tall, strong bodies for the sake of the Japanese nation (Frühstück 2003: 21–2).

Sumiko was not alone in marrying out. Flying for Pan Am placed these women on a global stage of desirability. The fact that this was a highly gendered moment was not lost upon them. In becoming stewardesses, these women accepted the sexualized and often racialized terms of their employment. Many of them married 'up' – in anthropological terms, hypergamy – including at least one case of a marriage to a (white) Pan Am executive. Others may have married laterally in terms of socioeconomic position, but by marrying a white American, their marriage could be interpreted as hypergamous – racially, nationally and culturally.

International stewardesses' practice of marrying up (and out) gained notoriety in a book published by Fumiko Takahashi, a former Pan Am stewardess, entitled *Gaikokujin Dansei to Tsukiai Hou* (How to Date a Foreign Man, 1989). Drawing upon her experiences of travelling globally and eventually marrying an American, Takahashi gives tips and cautionary notes about men in twenty-five different countries. In each case, Takahashi emphasizes meeting elite men – with an emphasis on white males, even though she includes various countries. Using her own experiences as a Pan Am stewardess, and especially one who ended up with the pot-of-gold American husband, gives her words authority and the book its validity. Note, too, that in contrast with other women I interviewed who have made their permanent homes outside Japan, Takahashi chose to live in Japan. She thus represents yet a differently nuanced cosmopolitanism – living in Japan but married to a white man – that finds resonance within the confines of the nation.

Most women I interviewed expressed appreciation for the company that gave them the opportunity to work the world. During the 1960s and 1970s when Japan's economic ascendancy seemed unstoppable, these women rejected not only Japanese men, but also the companies that employed them. They opted instead for an American company, at least a temporary American home, and for some, an American husband. They engaged in processes of what Robbins calls 'emergent cosmopolitanism' – that is '(re)attachment, multiple attachment or attachment at a distance' (1998: 3) – defining cosmopolitanism etched in Euro-American terms, rather than Japanese ones, moving to the far side of the horizon. In effect, their actions – elite, globalized, and ultimately defiant – allowed these Japanese women to reject the confines they were meant to inhabit. However, even those who chose to eventually and permanently reside in Japan, configured subjectivities anew in subtle forms of re-assemblage. Shielded by elite class positions, their cosmopolitanism challenged the Japanese status quo through the texture of their subsequent lifestyles. The accumulation of experiences, sights, tastes, sounds, and ongoing social relationships drawn from their work as stewardesses in the most prestigious international airline of the time became part of the fabric of their beings. They thus assumed a mantle of worldliness in ways and substance that most Japanese could not.

The women who flew for Pan Am constituted the 'Japanese in every jet' that became Pan Am's policy and goal. As the public face of the airline, they represented corporate cosmopolitanism – undoubtedly assisting Japanese passengers, but equally importantly, adding a new, exotic face and body to the roster of women in Pan Am's uniform. They joined Americans and Europeans and, at the time, a few Latin Americans. At the beginning of every flight, Pan Am's performance of its mastery of the globe included a greeting over the public address system by the different language speakers (and thus, countries and races) among its flight attendants. Adding Japanese to this line-up added immeasurably to the corporate bragging rights summed up as 'Pan Am's World'.

The particular spectacle of Japan as an emerging global power in the 1960s created the logical expectation that every jet might indeed carry a Japanese. That this expectation should rest on the shoulders of Japanese women to serve primarily, though not exclusively, male passengers speaks to the gendered nature of this cosmopolitanism. Both challenging and building upon old-fashioned images of Japanese femininity, the Japanese stewardesses with Pan Am gained celebrity as the modern girls of the time. Flying around the globe wearing the American uniform of the world's most prestigious carrier, they embodied the gendered, racialized, and class dimensions that defined the cosmopolitanism of the period. For the women themselves, Pan Am's world became their own, shaped initially within predictable power structures, but ultimately enacted with meanings and purposes of their making. Inhabiting the 'other side of the horizon' meant asserting Pan Am bodies and selves as part of their own emergent cosmopolitanism.

## References

Appadurai, A. (1996). *Modernity at Large: Cultural Dimensions of Globalization.* Minneapolis: University of Minnesota Press.

Arima, N. (2004). *Pan Am Kaisouroku.* Tokyo: Pan American Alumni Association.

Barry, K. M. (2007). *Femininity in Flight: A History of Flight Attendants*. Durham, NC: Duke University Press.
Bolen, M., K. Burton and F. van Hartesfeldt (prodrs) (1954–60). *Father Knows Best*. CBS (1954–5, 1958–60) and CBS (1955–8).
Fruhstucke, S. (2003). *Colonizing Sex: Sexology and Social Control in Modern Japan*. Berkeley: University of California Press.
Fukuda, J. (1963). *Hawaii no Waka Daishō*. Toho Studios.
Hannerz, U. (1990). 'Cosmopolitans and Locals in World Culture'. In M. Featherstone (ed.), *Global Culture: Nationalism, Globalization and Modernity*, pp. 237–52. London: Sage Press.
Hochschild, A. (1983). *The Managed Heart: Commercialization of Human Feeling*. Berkeley: University of California Press.
*Japan Times* (2005). 'Obituary: David Jones' (2 June). <http://search.japantimes.co.jp/cgi-bin/nn20050206b3.html> accessed 7 March 2008.
—— (2005). 'David Jones, Pan Am's Sumo Trophy Presenter, dies at 89'. <http://www.japantoday.com/jp/news/326904> accesssed 7 March 2008.
Kawauchi, K. (2004). *Pan Am Kaisouroku*. Tokyo: Pan American Alumni Association.
Kelsky, K. (2001). *Women on the Verge; Japanese Women, Western Dreams*. Durham, NC: Duke University Press.
Kurotani, S. (2005). *Home Away from Home; Japanese Corporate Wives in the United States*. Durham, NC: Duke University Press.
Massey, D. (1994). *Space, Place, and Gender*. Minneapolis: University of Minnesota Press.
Mills, C. W. (1951). *White Collar*. New York: Oxford University Press.
Nakata, H. (2001). 'Travel Show Star Believes Going Abroad Offers New Perspectives'. *Japan Times* (29 July). <http://search.japantimes.co.jp/print/nn20010729a4.html> accessed 8 March 2008.
*Life* (1967). 'Newest Stewardess Fad: A Japanese in Every Jet'. (Asia Edition) (1 May), 42–6.
*Nippon Times* (1955a). 'Japanese Hostesses'. (29 June), 6. [Pictorial caption]
—— (1955b). 'Nippon Hostesses Enjoy Big Demand'. (20 September), 6.
Otomo, R. (2007). 'Narratives, the Body and the 1964 Tokyo Olympics'. *Asian Studies Review* (June) 31: 117–32.
Owen, T. and W.S. Roberts (prodrs) (1958–66). *The Donna Reed Show*. ABC.
Pempel, T. J. (1998). *Regime Shift: Comparative Dynamics of the Japanese Political Economy*. Ithaca, NY: Cornell University Press.
Pratt, M. L. (1992). *Imperial Eyes: Travel Writing and Transculturation*. London: Routledge.

Robbins, B. (1998). 'Introduction, Part I: Actually Existing Cosmopolitanism'. In B. Robbins and P. Cheah (eds), *Cosmopolitics: Thinking and Feeling beyond the Nation*, pp. 1–19. Durham, NC: Duke University Press.
Sato, K. (2004). *Pan Am Kaisouroku*. Tokyo: Pan American Alumni Association.
Suzuki, M. (2004). *Pan Am Kaisouroku*. Tokyo: Pan American Alumni Association.
TBS (1959 to 1990). *Kanetaka Kaoru Sekai no Tabi*. Tokyo Broadcasting System Network.
Takahashi, F. (1989). *Gaikokujin Dansei to Tsukiau Hou*. Tokyo: Shufu to Seikatsusha.
Taurog, N. (1961). *Blue Hawaii*. Paramount Pictures.
Yano, C. R. (2011). *Airborne Dreams: Gender, Race, and Class in Post-war America*. Durham, NC: Duke University Press.
—— (2013). '"Flying Geisha": Japanese Stewardesses with Pan American World Airways.' In Alisa Freedman, Laura Miller, Christine Yano (eds), *Modern Girls on the Go: Gender, Mobility, Globalism, and Labor in Contemporary Japan*, pp. 85–106. Stanford: Stanford University Press.

JOY HENDRY

# Rewrapping the Message: Museums, Healing and Communicative Power

## Museums and Display

Museums and other forms of public display have been changing, within a global context, in order to accommodate the views of peoples whose cultural heritage is on show. In the colonial past, museums often formed part of the construction of national or imperial power and the people whose ancestral objects were chosen for display had little choice about their use. More recently, as ideas from one part of the world have spread rapidly to others, those whose ethnic groups have been showcased in museums have taken more of an active role (e.g. see Conaty 2003; Doxtator 1985; Eoe 1990; Fuller 1992; Hakiwai 1990; Hendry 2002; Kaplan 1994; Pannell 1994; Simpson 2001). Although local manifestations have been various, there are some common features which I examine here.

The themes I pick up here resonate quite well with those of globalization and assemblage, while the ones that are more specific to this paper are recuperation, or healing. By using the term global, I refer to communication and understanding that would appear to be shared by people who are widely separated geographically, but who have aspects of their history in common and who demonstrate the common features I will describe. By using the term violence I refer to the colonial appropriation of lands, peoples and customs, including their representation in museum display. I will try to show that an example of shared understanding is to be found in the notion of healing that I will analyse; a notion that also illustrates the power of the rewrapping of messages referred to in the paper's title (see Hendry 1993).

The notion of wrapping is used to describe ways of expressing meaning, subtly and often in layers. It can be applied to various sorts of non-textual communication in any society, notably in examples of cultural displays of power and influence, but it is particularly well developed in the Pacific communities where it is positively ineffective to speak too directly (e.g. Brenneis and Myers, 1984). In all cases wrapping is used strategically to create and express social situations and to manipulate relative positions of power. The relevant example here is the way that museum curators in former colonial situations have been rewrapping their displays to present and therefore express the continuing, but now more powerful, presence of the Indigenous people, once put on display as if they were disappearing, now often themselves in charge of the arrangements.

I begin with a place that has been subjected to several periods of violence throughout its long history, but which has also espoused an outward looking view of the world that has at the same time laid a good foundation for the global communication in which it is now very much involved. It has been recuperating from the effects of a sort of double colonization, first by Japan, and then by America as part of the Allied Occupation that followed the Second World War in that region. The place is Okinawa (the Ryūkū islands), site of some of the most terrible battles in which Japan engaged, location of a horrendous death rate due to the illness and poverty that ensued, but now known as a place of healing (*iyashi*), visited by people from wider Japan for its healthy food, its relaxed lifestyle, its spiritual richness, and its reputation for longevity (Prochaska 2013).[1]

To understand properly the theme of rewrapping requires a shift of focus – from the view conventionally accepted about a place by those who see it from the outside, or by those outsiders who inhabit the land, to one that better represents the view of those who are local, for whom it forms the centre of a worldview. In anthropological terminology, this would be a shift from an *etic* focus to an *emic* one, in this case including the casting of an *emic*

1 This paper is a version of a lecture I was invited to give in Oslo in March 2007 to open the programme of the Japan Anthropology Workshop meeting, which was organized by Arne Røkkum, one of JAWS' founding members.

view within Japan as *etic* from the perspective of Okinawans. In this way, the islands that make up Okinawa prefecture, usually regarded as so peripheral[2] to mainland Japan in that they run right off all the standard maps of the country, become the centre of a rewrapped world that looks very different.

Okinawa also makes a good choice for illustration of the concept of multiculturalism, with its own very real sense of *taminzoku* (multiculturalism) that goes beyond the fact that they themselves embrace a culture different from that of Japan's main islands. In short, the identity of the people who live in this region is based on a different history to that of the larger islands of Japan. Formerly a kingdom in its own right, with objects displayed in Shuri castle in Naha, the capital city, that demonstrate that fact, this is a region that has also skilfully negotiated long-term relations with many other of its neighbours, apart from the main islands of Japan. In fact, during my visit to Okinawa I was reminded more of another place which has clearly demonstrated its broad, global perspective, namely my home country of Scotland, than of anything peripheral at all ... and I did discover that a group of Okinawan lawyers are studying the Scottish parliament as a potential model for their own future autonomy.

A very neat material illustration of all these characteristics, along with the idea of message-rewrapping, is available to even the most casual of visitors to Okinawa's main island, if they take the trouble to visit the art museum in Urasoe City. There is but one permanent collection inside, and it is of lacquerware dating back to the fifteenth century. The English version of the museum brochure explains the choice as follows:

> Ryukyuan lacquerware is a radiant example of Okinawa's artistic heritage. This museum places much of its attention on this collection and on the cultural exchanges with neighboring countries that affected the artistic development of Ryukyuan lacquerware. We, at this museum, stress the importance that Okinawa has given, and gives, to 'looking outward' beyond the borders by which it is surrounded.
>
> (Urasoe Museum Brochure)

2 The area I chose as a first focus is in fact such an apparently 'marginal' place that at the national university there I met a Tokyo University anthropologist who had recently relocated precisely in order to set up a project on peripherality.

Before entering the galleries the exhibition begins with a literal piece of wrapping in the shape of a stunning six-part folding screen, depicting the trading port of the Ryūkyū Kingdom during this early period, seething with seafaring vessels of great variety, flying a similar diversity of ensign flags. This sets the scene for the display inside, which continues the theme of demonstrating Okinawa's overseas links. To give a few examples: a box with a design that combines a mother of pearl inlay in black lacquer said to characterize the Ryūkyū style reminiscent of rich pieces that were made under royal administration in the sixteenth century, is described as having been presented both to the Japanese shogunate and to the Chinese emperor. Another characteristic style, a red cinnabar piece with gold inlay, is presented to illustrate influences from China, but is then combined with mother-of-pearl, an indirect influence from Spain and Portugal in the Azuchi-Momoyama period (1568–1600). Ryūkyū style is said to have resisted influence from Japan, even after the invasion of the Satsuma clan in 1609 which enforced a *de facto* incorporation into the Japanese nation, and many designs continued instead to reflect Chinese-style landscapes.

Displays in the museum also illustrate lacquerware designs that were developed in other parts of Southeast Asia, such as Burma, the Philippines, Thailand and Vietnam, and explanations cover how these differed, as well as explaining the ways that the various influences travelled within this vast area. Trade with, and influence from, Korea was another theme, and a further piece displayed was cited as an example of work made specifically for export to Europe. Altogether the museum is definitely outward-looking, presenting lacquer from elsewhere and influences on local style. It is also historical, illustrating the various relationships of the Ryūkyū Kingdom; and it is very much concerned with local identity, presenting both distinctive techniques and an example of local resistance to their dominating near neighbour.

## The Rewrapping Theme

My recent research has actually been largely outside Japan, though concerned, as I have been here, with examining forms of cultural display and the messages they carry. In particular, I have been looking at the way some messages have been *rewrapped* by people who were peripheralized – like the residents of Okinawa prefecture – by a larger nation that subsumed them, sometimes even almost swallowing them up altogether. The results of my research have been largely positive messages of renewal, involving re-centring; recovering rights and the ownership over one's own messages; of re-presentation as self-presentation; and of reclaiming control over one's own identity, even sometimes one's very existence, where this had come under threat (Hendry 2005). Last, but by no means least, I learnt that these forms of renewal are a means of healing the wounds of past violence, an idea portrayed explicitly in cultural centres I visited in Canada, but implicit in plenty of others. I will consider some examples of all these, and then return finally to link Okinawa into the general pattern.

The other end of the Japanese nation, Hokkaido, provides a good start, for the case of the people known as the Ainu was one of the triggers for my outside research. Here is a very different situation to Okinawa, despite their shared peripherality, and the interpretation offered will be a little more complex and subtle. The broader context is not much presented in Ainu displays, indeed their sense of internal identity is still rather fragile (e.g. Hanazaki 1996; Siddle 1996), but it is a case in which considering the broader context will help provide an understanding of the rewrapping process that is taking place. I have the advantage of a long-term view here, and this I would like first to illustrate, inviting readers perhaps to make some interpretations of their own, though I will save mine until the chapter's end. Let me start by describing a couple of public places that seek positively to *re-present* (rather than rewrap) Ainu culture, both in the northernmost island of Hokkaido.

The first is known as Akan *kotan*, a community situated in the north east of the island, not far from the lake of the same name. I first visited this community in 1971, as a neophyte in Japan, essentially travelling as a tourist, following the trail of anything interesting. Here was an Ainu 'village', I was told, and the site consisted of a single street, with reconstructed houses that had been used by Ainu in the past, some containing items of material culture, in others sat people carving wooden bears, some with a salmon in their mouths. There was a live bear tethered in the middle of the street, and several young people walking about in garments I took to be Ainu clothing. The place was a little tumbledown, however, and when I sought to strike up a conversation with some of these 'Ainu', and ask them about their lives, they told me that the Ainu had basically died out and they were, themselves, students from Tokyo. In some ways, this statement about the Ainu was true, for the people who lived in this area had been subjected to such a stringent assimilation policy by the central government of Japan that their language and much of their distinctive culture had been virtually wiped out.[3] They had also been inculcated with strong disincentives ever to mention their Ainu origins.

I returned to this same street in 2004, and found the place completely transformed. First of all, most of the houses that fronted onto the street had been rebuilt in a variety of touristy styles, so there was a general air of commercialism that gave the place quite a different atmosphere. Now the people who worked in the shops, restaurants, and carving sheds, were happy to call themselves Ainu, indeed they were proud of the fact. There was a large new building reconstructed in an older style, but smart and fresh, housing a central internal hearth and a series of other examples of contemporary and historical Ainu material culture. There was also a large theatre, where daily performances included dancing, singing, haunting mouth-harp music, and the depiction of an Ainu legend. A more formal museum opened out onto the access road at the back of this street.

3 The Japanese policy with regard to the Ainu followed quite closely that of European colonial powers that were advised that the best way to deal with Native or Aboriginal peoples was to assimilate them into the wider society, see, for example, Hanazaki (1996) and Siddle (1996).

In the house of Ainu life (*seikatsukan*) I met Sigiko[4] Teshi, who not only declared herself to be Ainu, but was old enough, at 73, to explain the transformation that seemed to have taken place. She had lived in the area throughout the whole period, as had many of the other people working there, but back in the 1970s, most people did not want to put themselves on display, she explained. They were discriminated against as Ainu, so many were trying hard to pass as ordinary Japanese. Even those who resisted such pressure disapproved of using their apparently despised cultural heritage for touristic purposes. In the interim, contact with other Indigenous peoples had been gradually increasing, notably in Canada, and a small political resistance movement had been encouraged in a global context, coming to a particularly fruitful head in 1993, the year which was declared by the United Nations as the Year of Indigenous Peoples (http://www.un.org/esa/socdev/unpfii/). The Japanese government actually continued to reject Ainu demands for recognition as an Indigenous people until 2008, but it did pass a law for the protection of Ainu culture in 1997, and considerable resources have been invested in projects, such as museums, language revival and performance, which all contribute to the process of healing I mentioned earlier (see also Hendry 2005). The transformation of the village is an expression of the new pride in their cultural origins that the same healing makes possible.

The second site that proudly and happily represents Ainu culture is the Poroto-kotan in Shiraoi, not an excessively long journey from Sapporo, and close to the main railway line, so much more accessible for the casual visitor to Hokkaido. In fact this site was doing a good job in 1986, when I first went there and talked to those Ainu who were employed to dance, tell stories, and carry out other work within the site. At that time, parts of it were being constructed to portray directly the material culture of Ainu history, and I watched a couple of men building a canoe. It seemed a little anachronistic that they were using a chain-saw for the purpose, but when I asked them about it, they put me in my place: 'Why should we not take advantage of modern technology? We are Ainu, making an Ainu canoe,

4 This is an Ainu spelling of the name, which differs a little from the usual Japanese.

and these things are available for our use. Our ancestors would have done the same, had they had the resources.' This served as a reminder that they had by no means died out, and that they were making decisions about their own representations, whether or not they made sense to outside observers with their own expectations. As if that were not enough, the dancers, whom I asked if they considered themselves Japanese, as well as Ainu, replied without hesitation: 'Of course we are, we are the first Japanese.'

## The First Canadians

My research outside Japan was initially concerned with alternative epistemologies about conservation and exhibition,[5] and I visited culture centres, museums, tourist shows and communities in several different countries, where I discovered a real global movement of renewal and regeneration among people who call themselves Indigenous. In order to look at it in more depth, I spent a period of eight months in and out of a First Nations Reserve in central Canada, called the Woodland Cultural Center. This was founded in 1970 as part of a surge of support for healing past rifts[6] that had been displayed to the world by the Native people of Canada at the 1967 Montreal EXPO in a pavilion which they were invited to build by the Canadian government.

In the feasibility study for the Center, much was made of the need to heal the wounds of past iniquities suffered by their people, and of how

5 The results of this initial research are to be found in Hendry (2002).

6 One of the chief resentments among the First Nations in Canada was a system of residential schools that were used to implement the policy of assimilation, and the Woodland Cultural Center had been such a school, known as the Mohawk Institute. Children were sent there from quite distant parts of the country, punished for using their own languages, fed basic meal (known as mush) while their labour was used to grow fresh vegetables to sell to the local community, and rarely if ever were they allowed to return to their homes (e.g. Graham, 1997).

the cultural activities of such a healing process would help to solve problems such as poverty, high school dropout rates and alcohol dependency. The museum tells a story similar to that displayed at the 1967 Universal Exposition (EXPO '67). I had visited that pavilion and, nearly fifty years later, I can still recall its powerful images of disempowerment, despite the help offered by the natives to the new arrivals in their land, and the native influence in the construction of the new Canada. The Canadian government was shocked by the story, and its display to the world at large, and I believe its response helped to place the First Nations of Canada at the forefront of the future global developments. It was mentioned specifically by Ainu people as having helped them in their struggle for recognition.

A crucial part of being able to use one's own culture as a healing resource, taken completely for granted unless it is questioned, is that a people must be in control of their own representation. For Indigenous peoples around the world, their material culture has very often been collected, by anthropologists and other interested parties, and displayed in museums. Sometimes these are national museums, where the representation of local people has been a part of the nation building project; in other cases the objects were brought by travellers and researchers to distant locations where they have been set out as a kind of representation of imperial accomplishment, and used for research. Anthropologists, in general, might choose a more scientific way of describing the collections they have acquired and deposited in ethnographic museums, but in the end, Indigenous people rarely have been the ones who own and control these representations.

One of the discourses that is current among Indigenous peoples, then, is that they want at the very least to be consulted about collections of their peoples' artefacts in museums. There are various parts to this discourse. The most serious is that museums hold large collections of human remains, seen by contemporary Indigenous people as ancestors who cannot rest in peace, and much negotiation is taking place about their return (Leggett 1999; Mihesuah 2000; Peers and Brown 2003). An issue more relevant to this chapter, however, is that museum displays very often use the past tense in their descriptions of the people whose cultural artefacts they show (Doxtator 1988; Hendry 2005; Hill 2000). It might be argued by curators and designers of such displays that the cultural arrangements do indeed represent a situation now gone, but the descendants of those whose lives are depicted argue that it portrays them as if they had died out. Not only are they denied charge of their own displays, then, but such displays act to negate their continued existence.

More to the point, in recent years, there have been some serious changes in the way that museums around the world represent Indigenous people, and some of these changes propose that the cultural resources enshrined in the collections are ways of renewing identity and affirming a continuing creativity. At the Woodland Cultural Center, the museum is run entirely by members of the Haudenosaunee and Ojibwe Woodland peoples after whom it is named, and it does two important innovative things. First, it tells the story of the people, from pre-contact with European travellers, through trade and the gradual erosion of their traditional lives, to the present situation. Secondly, it has regular temporary exhibitions of contemporary art which demonstrate the talents of the living people.

Now – although it took until a crisis[7] in the later 1980s for this to be formalized – museums around the world are finding it necessary to consult the descendants of the people whose objects they have on display even if they retain overall control. In areas where Indigenous peoples reside, they are often involved in the design of new exhibitions, their voices may be heard, reproduced for the visitor in various ways, and it is quite usual for there to be displays of contemporary art work alongside the representations of earlier life. The new Museum of the American Indian that opened at the Smithsonian in Washington, DC in the autumn of 2004 is directed by a Native American, as was the National Museum of Australia by an Aboriginal

7 This crisis came about as the Glenbow Museum in Calgary prepared an exhibition to coincide with the Winter Olympics. Some of those whose material culture was to be represented wrote to museums around the world and asked them to boycott the exhibition due to the fact that one of its sponsors was the oil company, Shell, which had just been given the right to exploit land that the Lubicon Cree regarded as traditionally theirs. The situation was eventually resolved by the setting up of a Task Force which had equal representation from museums and First Nations and which published a list of pointers for a way forward which has been largely followed since (Phillips 1990; Task Force 1992).

woman, though unfortunately the three-year contract of the latter was not renewed. Te Papa, the national museum of New Zealand, has such a sophisticated Maori section that the Pakeha or white New Zealanders complain that they have been short-changed (Goldsmith 2003).

## Other Examples of Change

In locations more distant from the people themselves, exchange arrangements are made so that Indigenous people can visit the creations of their ancestors and even borrow them. A relevant example here is the relationship between the Royal Scottish Museum in Edinburgh and the Ainu people whose work was collected by the doctor/traveller Neil Gordon Munro. Objects have been exchanged for temporary displays in each country, and Ainu artists have travelled to Britain to demonstrate their skills as well as to examine the work of their ancestors. I met one such group, who came to do a display of Ainu dancing in Oxford, when I was asked to give them a tour of the Pitt Rivers Museum afterwards. I was, in return, ensured an excellent reception from one of the dancers, Masahiro Nomoto, when I re-visited in 2002 the Poroto-kotan at Shiraoi described above. A recent initiative involving the Japanese National History Museum's efforts to digitalize the materials of Neil Munro held in the UK has required careful negotiations with representatives of the community in Hokkaido where Munro lived and collected many of his Ainu materials.

This community in Nibutani can boast two Ainu culture museums, but each wraps their culture in rather different ways, conveying a message not immediately evident to the casual visitor. The older of the two looks rather old-fashioned, by modern museum standards, but it was put together almost entirely by a man who always maintained his Ainu identity, and indeed he represented his people as a member of the Japanese parliament. This is the late Shigeru Kayano, who also wrote many books about Ainu people and culture, and the objects in the museum were collected throughout his lifetime to protect them from the threat of extinction that hung over them for so long. The other museum has clearly had a greater investment in its construction, but it was built as part of a Japanese government compensation package offered to the local council in exchange for permission to divert much of the water from the Saru River, a chief source of the salmon that ensured the livelihood of Ainu people in that area, into a dam. Kayano and others opposed this project, but a majority on the council accepted it, so the two museums also stand as evidence of a split in the local community.

Some Indigenous people do not like museums at all, arguing that they always give an impression of a people passed on, and so they rewrap the messages they want to portray to the world in a variety of different forms. In Haida Gwaii (formerly known as the Queen Charlotte Islands), found just off the American Northwest coast, the local museum has been replaced – as has the outsiders' name for the isles – by a huge heritage centre with a performance space, meeting rooms for elders, eating areas, and a variety of different displays that seek to demonstrate local ideas. When I visited in 2003, they had raised several totem poles to mark the area for the construction – new poles constructed especially for the occasion, as Haida artists see no value in preserving old totem poles, as museums are wont to do, when they can make beautiful new ones that represent their continuing skills. The cultural healing that this kind of activity involves, they say, offers a parallel with the pride taken in the newly constructed old-style house I visited in Akan kotan, and the chainsaw made canoe in Shiraoi. In Nibutani, I found local council workers who had been given time off work to recreate the carving and sewing skills of their grandparents, and delightedly they showed me the garments they would now be able to wear with new-found pride.

Another Canadian example, in Vancouver Island, is the U'mista Cultural Center, built by the Kwakwa̱ka̱'wakw people at Alert Bay in order to gain the return of and to house masks that had been confiscated during the period when the strongest assimilation policies were imposed on them. An important part of their ritual and socioeconomic life had been the practice described by anthropologists as potlatch, meetings when vast quantities of goods were given away to mark a special occasion, and which were part of a continuing exchange of wealth. Towards the end of the nineteenth century, these meetings were banned by the Canadian government, under pressure from churches, among others, but so important were they to the local people that they carried on holding them in secret. On one terrible occasion in 1921, when they were caught, they had had to sacrifice their ceremonial masks in order to gain the release of their elders, who had taken responsibility for the events, were arrested and incarcerated. The masks were kept carefully in a museum, and when the Canadian government reversed their policies, they asked the Kwakwa̱ka̱'wakw to build a similarly protective home for them as a condition of their return. This they named *u'mista*, meaning homecoming, and the occasion became part of a great healing process, recorded in the displays, as well as in a couple of documentary films (Clifford 1991; Jonaitis 1991; Stanley 1998). In this centre, I also found a whole case devoted to Ainu materials, presented as part of an exchange that included school visits, performances, and ongoing communication, all made possible by the Japanese government's support for cultural preservation.

The Ainu may not have won control of their own lands, or even as much political power, as those other Northern peoples, the Inuit, have in the new Canadian province of Nunavut, but at least in places like Akan- and Poroto-kotan people who call themselves Ainu can use chainsaws if they want to, and rewrap their representations in their own Ainu way. On one visit I made to Shiraoi, I met a group of school children visiting from Okinawa, who were impressed by the display of Ainu identity, and who claimed to share a kindred spirit with these seemingly peripheral peoples at the other end of the nation that has subsumed them both. In both cases, there is much evidence of healing enclosed along with the forms of rewrapping we find in museums and more innovative examples of display.

Indeed, as mentioned above, the Okinawan atmosphere is now sought by people from wider Japan for its healing (*iyashi*) qualities, for its healthy food, its 'slow' lifestyle, and its reputation for longevity. The healing islands offer cultural elements like *kachashii* music, where all the people dance together, and sacred groves of the spirit world still regarded as close and influential in daily life. Okinawa is a popular destination for recovery from the fast life of mainland Japan, a place with resort hotels which offer wellness programmes, and a general atmosphere of comfort and care. Here, surely is an example of some kind of communicative power where the wrapping or rewrapping has spilled out beyond the walls of the museums.

## References


Assembly of First Nations and the Canadian Museums Association (1992). *Task Force Report on Museums and First Peoples*. Ottawa.

Brenneis, D. and F. R. Myers (eds) (1984). *Dangerous Words: Language and Politics in the Pacific*. New York: New York University Press.

Clifford, J. (1991). 'Four Northwest Coast Museums'. In I. Karp and S. D. Lavine (eds), *Exhibiting Cultures: The Poetics and Politics of Museum Display*, pp. 212–54. Washington, DC: Smithsonian Institution Press.

Conaty, G. (2003). 'Glenbow's Blackfoot Gallery: Working Towards Co-existence'. In L. Peers and A. Brown (eds), *Museums and Source Communities: A Routledge Reader*, pp. 227–41. London: Routledge.

Doxtator, D. (1985). 'The Idea of the Indian and the Development of Iroquoian Museums'. *Museum Quarterly* (Summer 1985): 20–6.

—— (1988). *Fluffs and Feathers: An Exhibit on the Symbols of Indianness; a Resource Guide*. Brantford, Ontario: Woodland Cultural Center.

Eoe, S. (1990). 'The Role of Museums in the Pacific: Change or Die'. *Museum* XLII (1): 29–30.

Fuller, N. J. (1992). 'The Museum as a Vehicle for Community Empowerment: The Ak-Chin Indian Community Ecomuseum Project'. In I. Karp, C. M. Kreamer and S. D. Lavine (eds), *Museums and Communities: The Politics of Public Culture*, pp. 327–66. Washington and London: Smithsonian Institution Press.

Goldsmith, M. (2003). '"Our Place" in New Zealand Culture: How the Museum of New Zealand Constructs Biculturalism'. *Ethnologies Comparées* (Printemps) 6 <http://alor.univ-montp3.fr/cerce/revue.htm>.

Graham, E. (1997). *The Mush Hole: Life at Two Indian Residential Schools*. Waterloo: Heffle Publishing.

Hakiwai, A. (1990). 'Once Again the Light of Day? Museums and Maori Culture in New Zealand'. *Museum* XLII (1): 35–8.

Hanazaki, K. (1996). 'Ainu Moshir and Yaponesia: Ainu and Okinawan Identities in Contemporary Japan'. In D. Denoon, M. Hudson, G. McCormack and

T. Morris-Suzuki (eds) *Multicultural Japan: Paleolithic to Postmodern*, pp. 117–32. Cambridge: CUP.

Hendry, J. (1983). *Wrapping Culture: Politeness, Presentation and Power in Japan and Other Societies*. Oxford: Clarendon Press.

—— (2002). 'Being Ourselves For Us: Some Transformative Indigenous Ideas Of Ethnographic Display'. *Journal of Museum Ethnography*, 14: 24–37.

—— (2005). *Reclaiming Culture: Indigenous People and Self-Representation*, New York: Palgrave.

Hill, R. W. (2000). 'The Museum Indian: Still Frozen in Time and Mind.' *Museum News*, 79(3): 40–4.

Jonaitis, A. (ed.) (1991). *Chiefly Feasts: The Enduring Kwakiutl Potlatch*. New York: American Museum of Natural History.

Kaplan, F. S. (ed.) (1994). *Museums and the Making of 'Ourselves': The Role of Objects in National Identity*. London: Leicester University Press.

Leggett, J. (1999). *Restitution and Repatriation: Guidelines for Good Practice*. London: Museums and Galleries Commission.

Mihesuah, D. A. (ed.) (2000). *Repatriation Reader: Who Owns American Indian Remains?* Lincoln, NE: University of Nebraska Press.

Pannell, S. (1994). 'Mabo and Museums: the Indigenous (re)Appropriation of Indigenous Things'. *Oceania*, 65: 18–39.

Peers, L. and A. Brown. (eds) (2003). *Museums and Source Communities: A Routledge Reader*, London: Routledge.

Phillips, R. (1990). 'The Public Relations Wrap: What we can learn from The Spirit Sings'. *Inuit Art Quarterly* (Spring): 13–21.

Prochaska, I. (2013). *Kamichu-Spirituelle Heilerinnen in Okinawa*. Wien: Praesens Verlag.

Røkkum, A. (2006). *Nature, Ritual and Society in Japan's Ryukyu Islands*. London: Routledge, JAWS series.

Siddle, R. (1996). *Race, Resistance and the Ainu of Japan*. London: Routledge.

Simpson, M. G. (1996, rev. ed.). *Making Representations: Museums in the Post-Colonial Era*. London: Routledge.

Stanley, N. (1998). *Being Ourselves for You; the Global Display of Cultures*. London: Middlesex University Press.

West, R.W. et al. (2005). *The Changing Presentation of the American Indian: Museums and Native Cultures*. The National Museum of the Native American and Washington University.

# PART II

# Contemporary Configurations

WOLFRAM MANZENREITER AND JOHN HORNE

# Football in the Community: Global Culture, Local Needs and Diversity in Japan

## Introduction

This chapter explores ideas about community in contemporary Japan and different means of community development in relation to football's globalization. Some of our previous writings on football in Japan have already attended to these issues, in particular our reading of football in Japan as an 'ideological soccer apparatus' (Horne and Manzenreiter 2008) and our study that was the first to consider the 'footballization of East Asia' against the background of world systems theory (Manzenreiter and Horne 2007). For the purposes of this volume we have drawn upon these articles; updating them to take into account developments up to 2013. Asking what kinds of communities are emerging and making use of football for their specific objectives, we refocus on globalization and its local rootedness as basic aspects underlying the formation of distinctive football communities in contemporary Japan.[1]

In the study of community in the global age sociologists now deemphasize the centrality of place in the formation of identity, turning towards belonging, interest or attachment as alternatives, thus opening out the conceptual space within which non-place forms of community can be analysed (Castells 1996). However, the successful introduction of football to Japan depended on the deterritorialized production of football mediascapes as much as on the redefinition of localities, strengthened bonds of local affiliation and hence the continuous working of what some might

1 Whilst the main focus is on developments to the mid-2000s, a brief update on the condition of football in Japan in 2013 is discussed elsewhere.

think of as an antiquated sense of place. Given that over the past twenty years diverse cities and regions in Japan have been differently prepared to employ the game for their own local affairs, they have produced varying modes of community support and exhibited distinct types of football communities. Our research offers a taxonomy of football communities in Japan, demonstrating how football and its associated senses of locality have been used to cope with the emotional, social and economic challenges of the social change triggered by globalization.

The meaning of football in the making and remaking of communities has been the focus of both academic and political debate – at least in continental Europe. For various reasons that include local and ethnic identity, class and social change, football has acquired the characteristics of a representative sport through a long process that began during the social transformations of the nineteenth century. European football clubs, professional or amateur, came to represent, first, geographical locations, as their roots dated back to times when the entire life world of a community was largely restricted to the territory its members inhabited. Technological and cultural changes have expanded the frontiers of such life worlds; and mobility – physical as well as social, individual as well as collective – has largely increased, yet football clubs have persisted as a meaningful site for ritualized identification with a specific locality, a social group associated with a certain territory, or as an iconic image deriving from a combination of place, class and style. Today, whilst its supporters or players need not be restricted to the same geographical area (Bale 2000), it is a club's long association with a locality, a shared history and a common sense of belonging that lead policy makers to recognize football's potential effectiveness for doing the work of community development – either as a tool for community mobilization or accommodation to wider governmental programmes (cf. Football Task Force 1999; DTP 2006).

The Japanese football-community nexus differs from the European experience in three major respects. Firstly, football has played no significant role within Japanese communal affairs for most of its hundred-plus year history there (cf. Horne 2002). Until recently, it could not have done so because football had never attracted a large enough base of followers, either as a spectators or as a participants, before the Japanese Professional Football League's (hereafter, J-League) inauguration in 1993. Previously, clubs usually had represented a corporation, or sometimes a brand name, and were meant either to promote the corporate image or to provide their employees with a focus for identification (Sawano 2005). The spread of professional football, however, is closely linked to the main drivers behind the contemporary form of globalization: commercialization, financialization, media intensification and vertical as well as horizontal consolidation of transnational business activities (see Millward, 2011). Having reached Japan in the late twentieth century, the burgeoning popularity of the global game indicated both the successful integration of the football periphery into global commodity markets, as well as the transfer of relations of consumption in spheres where marketization previously had been close to non-existent in Japan, such as community relations.

A second distinctive feature of Japanese social development has been the relative stability of community patterns, which have proven far more resistant to the destabilizing effects familiarly seen in modern urban western societies. The prevalence of formal and informal institutions that organized communal life, often at the expense of individuals' desires, is not simply a continuation of traditional practices: sociological and anthropological studies (Fukutake 1989, Bestor 1989) have provided ample evidence that the so-called characteristic patterns of Japanese social organization were the result of conscious, concerted efforts to adapt traditional small-scale rural settlements and urban neighbourhoods to modern associational forms by finding a balance between expressive solidarity and instrumental association – or in Durkheimian terms, between mechanic and organic solidarity.

Rather than urbanization, it has been the move toward a technologically sophisticated and service-oriented, postmodern, consumer society that has arguably undermined the fundamentals of Japanese community patterns. Participation in communal affairs became segmented and selective, dependent on the capacities of the community to respond to the particular interests of specific residents (Lützeler and Ben-Ari 2005; Robertson 1991).

Additionally over the past three decades a moral panic has played out in the mass media about an increasing number of social anomalies, such as *ijime* (bullying), *enjo kōsai* (compensated dating), *oyaji bataki* (youngsters beating up old men), *parasaito shinguru* (unmarried adults living with and at the expense of their parents) and *hikikomori* (deliberate withdrawal from social life). These have suggested a diminution of the values and institutions that once integrated the individual into their community and regulated their life in this collective-oriented society (White 2002).

As with the populist discourses on communitarianism and new urbanism in the United States (Harvey 1997), in Japan, too, there exists the tendency to overrate the ability of the idealized community to counter threats to the contemporary social order and to understate its darker side. Such organizations were one of the key sites of social control and surveillance, bordering on overt social repression, which was especially the case in early modern and wartime Japan (cf. Fukutake 1989). To fight the flaws of late modernity with the evils of earlier social formations could be seen as a post-modern irony – yet the notion of community, revived or newly constructed, remains at the centre of contemporary debates about town making (*machi-zukuri*) in Japan; and quite a few advocates speak in favour of football as a means through which to do this.

This is less surprising if we consider the third difference between Japan's and Europe's vision of football communities. We will demonstrate in the rest of this chapter that throughout its entire, albeit relatively brief, history professional football in Japan has been closely related to predominantly top-down projects of urban regeneration and community development (cf. Manzenreiter and Horne 2005). In contemporary Japanese discourses, the organization and institutional practices that promote football and *komyunitii* (community) are best understood as a reaction to globalization's 'perils'; and a number of key variables are responsible for the emergence of distinct patterns of communal football support. Additionally, whilst much of the discourse on football and the Japanese community serves very different interests, it generally continues to facilitate the proliferation of private consumption, public spending and transnational money flows. Hence the football-community nexus in Japan inculcates amongst community members specific ways of understanding and shaping the relationship between the individual and society.

In previous publications we have demonstrated the importance of common interests and emotional attachment for the formation of a fan culture. Hence we are aware of the socially integrative effects of football for sub-cultural groupings (Manzenreiter 2006) as well as for more part-time members of ephemeral, otherwise dissociated, makeshift communities of event culture (Manzenreiter 2004a). Here we focus on football communities as groups of private and corporate residents sharing a particular local area that roughly corresponds with the geographic boundaries of aggregated support for the local professional team, because of the important role that place and territory continue to play for football fandom, even in the most advanced of information societies, such as Japan.

## The J-League's Corporate Community Model (CCM)

The J-League was incorporated as an autonomous, non-profit-making organization in 1991. Aside from promoting competitive football prowess, the J-League required each applicant club to be a registered corporation specializing in football, a stipulation designed to force the management of each club, as well as players and coaches, to be fully professional. Since J-League football, unlike semi-professional corporate football and professional baseball, was not to be used simply as a promotional tool, clubs were also requested not to have their owners' name as the team name. Hence the previous amateur side Toyo Industrials became the professional club Sanfrecce Hiroshima, Matsushita Electrics became Gamba Osaka, and Mitsubishi Heavy Industries changed their name to the Urawa Red Diamonds. In some cases the links with the past remain quite explicit. Jubilo Iwata's logo includes the words 'Yamaha FC', referring to the origins of the side as a company team; similarly, the Urawa Reds team is owned by the Mitsubishi Motors Football Club; both are independent corporations, benefiting from good relations with their former owner company, which now acts as a major, but not the sole, sponsor.

Since 1991 the J-League has gone through broadly three transformative phases, during which expansion has occurred, league arrangements have been modified and new clubs allowed to join. Forty professional teams have been established, dozens of large stadiums erected, and thousands of players, coaches, match officials and full-time staff have registered with the

Japan Football Association (JFA). In 2013, there were forty full members in the two J-League Divisions J1 and J2, eighteen in the top division and twenty-two in the second rank (see Table 1).[2]

Table 1. Professional Football Clubs in the J-League Divisions 1 and 2 in 2013

| J1 Teams = 18 | J-League (Year joined) | J2 Teams = 22 | J-League (Year joined) |
|---|---|---|---|
| Vegalta Sendai | 1999 | Consadole Sapporo | 1998 |
| Kashima Antlers | 1991 | Montedio Yamagata | 1999 |
| Urawa Red Diamonds | 1991 | Mito Hollyhock | 2000 |
| Omiya Ardija | 1999 | Thespa Kusatsu Gunma | 2005 |
| Kashiwa Reysol | 1995 | Yokohama FC | 2001 |
| FC Tokyo | 1999 | FC Gifu | 2008 |
| Kawasaki Frontale | 1997 | Tokushima Vortis | 2005 |
| Yokohama F. Marinos | 1991 | Ehime FC | 2006 |
| Albirex Niigata | 1999 | Avispa Fukuoka | 1996 |
| Shimizu S-Pulse | 1991 | Roasso Kumamoto | 2008 |
| Jubilo Iwata | 1994 | JEF United Ichihara Chiba | 1991 |
| Nagoya Grampus Eight | 1991 | Tokyo Verdy 1969 | 1991 |
| Oita Trinita | 1999 | Kyoto Sanga FC | 1995 |
| Shonan Bellmare | 1999 | Gamba Osaka | 1991 |
| Ventforet Kofu | 1999 | Vissel Kobe | 1999 |
| Cerezo Osaka | 1996 | Matsumoto Yamaga FC | 2012 |
| Sanfrecce Hiroshima | 1991 | Tochigi SC | 2009 |
| Sagan Tosu | 1999 | Kataller Toyama | 2009 |
| | | Gainare Tottori | 2011 |
| | | Fagiano Okayama | 2009 |
| | | Giravanz Kitakyushu | 2010 |
| | | V-Varen Nagasaki | 2013 |

Source: J-League Official Website <http://www.j-league.or.jp/eng/> accessed 8 January 2013.

2 These two professional leagues are at the top of a pyramid structure under which a nationwide amateur league, nine regional leagues and forty-seven prefectural leagues operate. The J-League plans to expand professional football to a Third Division.

While European social scientists have been concerned with the role football has played in the establishment of communal bonds, in Japan's case the question might be better inverted to ask: what has been the role of communities in professional football's development? Before the J-League, semi-professional and intercollegiate football had long existed in Japan yet it was less successful in terms of spectator turnout. Raising the performance of the national team as well as encouraging popular interest in the game – two central objectives associated with the J-League mission – inevitably required the professionalization of the sport's every aspect. Japanese capital, intercorporate networks and marketing know-how have been involved in the relatively recent commercialization of sport on a global scale (Manzenreiter and Horne 2002:10), reflecting the fact that Japan's football bureaucrats knew that commercialization was key to generating the funds necessary to turn a minority sport into a successful, viable product. Capital for new investment opportunities was ample at the onset of the 1990s, just before Japan's recession. Moreover, building up a loyal customer base, particularly in Japan's well-known fickle consumer markets, required a long-term vision, time and a strategic business plan. To bridge the first ten years of the new football market, a network of investors and stakeholders willing to share the risks of advance financing had to be developed (Hirose 2004).

Communities were assigned a crucial role in both the short and long term perspectives. Choosing from various models of sport financing, the J-League opted for an amalgam of the North American franchise system and the European sports club system (Ubukata 1994: 21, Horne and Jary 1994). The J-League copied the art of merchandising meticulously from the NFL (American Gridiron Football). Sony Creative Productions tailored a standardized set of corporate images for the starting line-up of ten professional football teams that were meant to appeal to young women, currently Japan's most powerful consumers (cf. Watts 1998). The franchise system's closed-club style guaranteed accounting control to the non-profit J-League organization over its member clubs' books as well as over central revenue streams – income from broadcasting rights, merchandizing and sponsorship is distributed to all member clubs, after deducting the funds necessary to cover operational expenses. The cartel-like organization safeguarded the teams, at least during the start-up period, against the competitive and economic dangers of relegation.[3]

The European sports club model, with its grassroots approach and emphasis on community service, became influential in promoting Japanese football for three reasons. Firstly, despite the American influence on many forms of contemporary Japanese popular culture, association football was neither regarded as a mass sport nor as a commercial success in the United States, as it was in Europe. Although soccer is now the largest youth team school sport in the United States, creating community bonds and attracting corporate sponsorship, in the late 1980s lessons from the United States about launching football were not considered useful. Secondly, the driving forces among the advocates and managers of the professionalization initiative came from a generation of ex-footballers who had been socialized into the sport during the 1960s when the German Dettmar Cramer was involved in coaching the Japanese national team. During occasional visits to football schools in Germany, young Japanese players experienced the excellent training facilities, available to amateurs as well as professionals, and the local population's enthusiastic mass support. The heart-warming image of 'sports for all'[4] in combination with the organization of top-level sports formed a lasting impression in the minds of the ageing Japanese football bureaucrats. This is very likely a major reason why Germany's sport and football clubs were explicitly chosen as the main role models for Japan's first fully professional football league, despite the multiple problems the European sports club model has faced since the 1960s.

Thirdly, communities as a focal point for club membership and as sponsors of club activities were urgently needed for J-League's success. In fact, the J-League demanded that any aspiring member club demonstrate that their hometown was willing to promote football in the region by

3 These two professional leagues are at the top of a pyramid structure under which a nationwide amateur league, nine regional leagues and forty-seven prefectural leagues operate. The J-League plans to expand professional football to a Third Division.

4 'Sports for All' was the core notion of the 'Golden Plan', a programme to improve the sports infrastructure and to grow sport participation rates by the German government in the 1960s.

providing financial guarantees and infrastructure projects to build stadiums and other facilities if needed alongside other more direct forms of capital investment. In exchange, J-League member clubs were required to 'unite with the community, familiarize people with the sport-oriented lifestyle, and contribute to the physical and mental well-being and pleasure of local society', as the official mission statement declares. To facilitate integration and identification, all J-League clubs were asked to forge names combining the geographical place name with a nickname of a particular local flavour. Resonant keywords for the J-League's corporate community model include the term hometown (*hōmu taun*), where all stakeholders are based, as a vernacular version of *oraga machi* (our town); and *chi'iki mitchaku* (regional adherence), meaning simply having a very close relationship with the hometown region. These are core notions to be found in virtually all public statements issued by the J-League concerning its mission. In 1996 when the J-League had lost some of its initial dynamism, it reemphasized its commitment to the public, particularly to the inhabitants of hometown areas, by the public expounding the J-Mission, or Centennial Plan (*hyakunen kōsō*), referring both to the 100-year history of European sport clubs and to the long term prospects of contemporary football initiatives in Japan.

Another prominent key phrase is *sanmi ittai* (trinity), referring to the ideal image of football in the community, based on the cooperation of civil society (*shimin*, townspeople) with businesses (*kigyō*) and local regional bureaucrats (*gyōsei*). The J-League serves as a hub and communication point between the dispersed actors in the region, channelling public funds into the football community. Popular interest in the local team is therefore crucially important to the team's commercial viability. Spectator turnout feeds directly and indirectly into club revenues, since ticket sales are a major source of income and attendance rates are a striking argument for the corporate community's sponsorship activities. Football clubs also cater to the needs of urban redevelopment programmes since they promise to enhance locals' quality of life in the region by providing healthy entertainment (*kenzen-na goraku*) as well as being a source of local pride and communal identification, unlike the earlier cultural and sporting hierarchy, which was largely restricted to the Tokyo conurbation in the east and the area around Osaka in western Japan. It was clearly hoped that these and other intangible benefits of the J-League would be effective in discouraging the younger generation from migrating to the large cities, while bridging the gap between older residents and newcomers to their neighbourhoods.

According to the J-League community model, football is the point of entry into a vibrant community life sustained by actors from the three different fields of business, politics, and civil society. While having their own particular interests, needs, and potentials, they also provide the three pillars that sustain football.

Football itself should be added as the fourth and distinctive player within the ideal new community, not least because the managing body of the J-League proposed establishing professional football throughout Japan for purposes far beyond the limits of the game. Its stated mission is: 'to foster the development of Japan's sporting culture, to assist in the healthy mental and physical growth of Japanese people, and to contribute to international friendship and exchange.' Such statements reveal a functionalist approach to the ideology of football, though they do not show whose interests it ultimately serves.

## The Making of Football Communities in Japan: Phases 1 and 2

From a macro-perspective, the instalment of professional football as an industry has been a remarkable success in the otherwise dark period of Japan's long recession. As noted, the league has expanded from ten teams in 1993 to forty competing in two divisions (J1 and J2) in 2013. As with any other market, it has gone through cycles and phases. Between 1993 and 1995 football boomed, attracting more than six million visitors in an average season and generating an annual turnover of ¥10 billion (~£50 million) for the J-League. But the hype did not last. During the period when teams constantly and rapidly expanded, average crowd sizes declined. In the 1997 season average attendances dropped close to 10,000, down from nearly 20,000 recorded three years earlier. Facing declining revenues from the turnstile and merchandizing sales, most, if not all, clubs were in the red.

Yokohama AS Flugels' collapse in 1999 alarmed the J-League and compelled them to seek a more reasonable and transparent style of club management from its members. With the national team qualifying for the 1998 World Cup in France and the 2002 World Cup – co-hosted with South Korea – looming, football came back into fashion. Since 2001 average attendances for first division (J1) matches have risen almost to their previous heights. With averages being close to 19,000 from 2004 through 2007 Japanese and foreign observers agree that football has come to stay in Japan. According to the analyst Hirose (2004), professional football created a market volume of ¥530 billion during its first decade. He also calculates that the J-League itself contributed only 0.1 per cent of the entire start-up investment: local authorities and communities contributed approximately ¥58 billion of the funds needed for infrastructure building, team development, marketing, etc. Particularly in terms of ownership, the joint efforts of local authorities, citizens and companies from the football teams' regions introduced a promising new sport business model, which is in marked contrast to traditional Japanese arrangements in professional baseball or contemporary sports business models prevalent in the United States and Europe. Nevertheless there exist tensions between the top down and bottom up approaches to community development and its association with football. A core tension is that of becoming international by being involved in a global game, while promoting a sort of traditional Japaneseness, as the following case studies will illustrate.

## Community Building Top Down: The Corporate Approach

The top-down approach dominated the making of football communities during its establishment and through the first years of the J-League's operation (circa 1991 to 1996). This era witnessed the league's enormous success, its gradual expansion from ten to fourteen teams and the confirmation that Japan would be co-hosting the 2002 World Cup. Although the

Japanese economy was visibly cooling down, there was as yet no sign of a severe crisis. The number of commonalties in the hometown images held by this first generation of professional football teams hinted at the way in which these had apparently been drafted on the drawing table. One of the driving forces behind the community building approach was the J-League itself. It acted as a football matchmaker, identifying and associating pockets of football-interested areas with clubs. Of the ten teams that started the J-League, nine were selected from former company sports teams that continued to benefit from a close relationship with their former owner company. Acting now as main sponsor, or major shareholder, of the company stock, company subsidies provided the completely inexperienced football management staff, often dispatched from the former owner companies' middle management, with a contract guaranteeing support against mid-term financial difficulties.

Virtually all of the first phase J-League clubs were founded within the densely populated areas of Japan's main island Honshu, which provided the largest spectator and customer bases. Tokyo was deliberately denied hometown status because the J-League wanted to avoid a confrontation between representatives of the capital and all other teams. Such a configuration – notably through the Yomiuri media corporation's control of the Tokyo Giants – had tainted power relations, the distribution of market values, revenue streams and spectator support during fifty years of professional baseball. Until 1995, the four new teams that joined the J-League, with the exception of traditional football powerhouse Sanfrecce Hiroshima, were settled in the Pacific Belt stretching between Osaka in the West and Kashima city in the East.

Kashima, with a population of 45,000, proved the exception to the rules laid out by the J-League for its prospective member teams in those years. In order to guarantee full capacity crowds on a regular basis, a hometown population of less than 100,000 was not regarded as large enough to fill the required 15,000-seater home stadium regularly. Hence the application of Kashima to be listed among the J-League hometowns, was initially considered to have little chance of success. However Kashima badly needed an effective re-imaging strategy. Like so many other towns and villages on the peripheries, Kashima suffered from the effects of failed land use planning. Located 150 kilometres from Tokyo and close to the Pacific Sea, until the early 1960s Kashima's inhabitants had been mostly agriculturists and fishermen. In the high growth era the greater Kashima area had become a specially designated industrial zone, attracting Sumitomo Steel, numerous component suppliers and more than 170 new companies into the area. Within two decades, the population nearly doubled. Numerous apartment buildings were constructed, a commercial harbour was opened, but in terms of amenities, social services and leisure programmes, the town could not keep pace with the rapid population growth. Given the conditions of limited quality of living and rising real estate prices, Sumitomo Steel, Asahi Glass and others faced difficulties in hiring enough workers, since neither company workers nor the young generation intended to stay permanently. Being part of the new professional football league promised to make a difference for Kashima.

Without the support of the industrial giant Sumitomo, which happened to own a company football team playing in a minor regional amateur league, the J-League would undoubtedly have turned down the application. Sumitomo Steel took the initiative, seeking support among local bureaucrats and business leaders while integrating the surrounding municipalities into its bidding campaign. Besides Sumitomo Steel and forty other companies from the region, five small local authorities became major shareholders of the incorporated Kashima Antlers Football Club. The J-League's main requirement, a roofed stadium with a capacity of 15,000 spectators (one seat for every third Kashima resident), was built largely at the expense of the Ibaraki prefecture government, which covered no less than 80 per cent of construction costs, estimated at ¥10 billion (Kubotani 1994: 50). In addition, public money granted by local authorities was used for renovating the traffic infrastructure. A twenty-year-old dream was realized when a branch railway line connecting Kashima to the Tokyo-Sendai track of the high speed Shinkansen train was opened. Additionally, public money financed the refashioning of a freight depot into a commuter station, the construction of parking lots, and improvements to sanitary and accommodation facilities (Koiwai 1994: 62ff).

All these investments were justified by Kashima Antlers' unexpected success on the pitch. In their inaugural season (1993) they won the first stage of the league only to be defeated in the final Suntory Championship playoff against the winner of the second stage, traditional football powerhouse Verdy Kawasaki (now Tokyo Verdy 1969). This remarkable performance soon sparked interest among local residents. Within months the official Antlers' fan club increased from just three to more than three thousand members. They were cheering for a team stuffed with Brazilian talent, including former world footballer of the year, Zico, who went on to manage Japan's national team between the 2002 and 2006 World Cups. Sumitomo Steel not only compensated for the huge deficits due to the foreign star players' high salaries, but also helped the Antlers to defray the clubhouse and training grounds' construction costs. Although reliance on a mother company was out of step with the official J-League community ideology, such subsidies were officially declared as advertising costs, rather than sponsorship expenses that would have attracted much higher taxation (Ubukata 1994: 52ff; 113f).

Thanks to this financial support, the presumed underdog team turned into one of the dominant club sides of the J-League in the 1990s and early 2000s. Winning the overall championship in 2001 provided the Antlers with their fourth title. The previous season they had won all three domestic competitions. Only Jubilo Iwata, another regular contender for the championship that enjoys the generous support of Yamaha Motors, was able to stop Kashima in 2002. The JFA acknowledged Kashima's leading role in the development of professional football by nominating the city as one of the ten 2002 FIFA World Cup host venues. When the designated World Cup venue was adapted to meet FIFA requirements, the investment of a further ¥23.6 billion in the stadium provided enough seats for almost every citizen in Kashima. The name of Kashima has become famous throughout Japan and Kashima itself a favourite object of study for town planners and local government administrations.

## Community Building Top Down: The Bureaucratic Approach

The success of the J-League tempted numerous communities to emulate the Kashima experiment. The J-League's second phase expansion took place from 1995, just as the Japanese economy was badly hit by a series of recessive periods and by the Asian financial crisis. Well into the following decade, there was low growth in productivity, real estate markets collapsed, unemployment rose, corporations went bankrupt, public debt mounted and the economy stagnated. Since the combination of economic growth and concomitant expectations of social security had come to play a crucial role in the Japanese collective image, forming a kind of economic nationalism, the loss of Japan's competitive edge called into question commonly shared ideas about what it meant to be Japanese.

Football provided potential solutions to these dilemmas. First, it afforded a useful metaphor for a new Japan. Rooting for a Japanese team became a new opportunity for showing national pride at international matches, particularly since the national team had strengthened and gradually came to draw even with South Korea. When Japan qualified for the 1998 World Cup in France, interest in the tournament became greater than ever, producing some of the highest television audience rates in Japanese broadcasting history. Second, as an integrative sport, the continuing presence of football in the Japanese regions provided a local alternative to a national collective identity. Third, as discussed above, the football industry also experienced the repercussions of the recessive climate, leading to the first – and so far only – closure of a team.

It is revealing of the ideological soccer apparatus that the J-League administrators explained the failure of AS Flugels Yokohama as due to a lack of interest in the team and lack of unity amongst supporters, rather than because market effects, sponsorship policies or the economic turndown in Japan had contributed to the decisions of its two main sponsors, airline company ANA and construction company Sato, to withdraw their support. When the forced merger with the other team based in Yokohama (the Marinos) was announced, fans banded together and petitioned the J-League about the dissolution of their club.

Despite this closure the J-League continued to expand. In 1999, a second professional football division (J2) was launched, increasing the speed of expansion. Hosting half of the 2002 World Cup was crucial during this period since more professional teams were needed in order to raise awareness of the game and its global flagship event, particularly in each of the ten designated host cities. Nearly one in two of the new thirteen J-League hometowns that came into existence during this phase was a potential future World Cup host city. In geographical terms this project-driven expansion of the infrastructure contributed to the further spread of football throughout the country, which was also an important step towards the formation of the imagined national football community needed for the World Cup fiesta. Professional football reached Japan's most northern island Hokkaido as well as the southern island of Kyushu for the first time. Football's spread created a new mapping of Japanese communities and of Japan itself.

In Sapporo Japan's most sophisticated roofed arena was opened. Giant stadiums with seating capacities of over 40,000 were also constructed in the hinterland regions of Tohoku in the Northeast, the backside of Japan facing the Japan Sea (Niigata), and in the main southern island of Kyushu (Ōita). Three more World Cup size stadiums were opened in large cities in Kanto and Kansai, when Cerezo Osaka, Kobe Vissel and FC Tokyo joined the league. Only three of the new teams – FC Tokyo (Tokyo Gas), Kawasaki Frontale (Fujitsu) and Omiya Ardija (NTT) – were direct descendants of former company sides, but even these could not entirely rely on the goodwill of a major company sponsor. Trust in club management was built on a new kind of ownership model in which local communities were deeply involved. In all instances, a consortium of some dozen companies and organizations, often including local authorities, constituted the team's formal owners. Another central feature of this period was the leading role played by bureaucracy and local elites in football community building. Two good, albeit quite different, examples of this development are the teams Ōita Trinita and Niigata Albirex.

Ōita is a small city (by Japanese standards) with a population of about 650,000. It is the capital of the homonymous prefecture, situated in Kyushu. For geographical reasons the terrain had impeded the development of industry and the integration of the region into larger communication and trade networks. Ōita was once famous for being the most depopulated prefecture in Japan (Arimoto 2004: 68), but it is also renown for the way football has been used by the local political elites to stem rural depopulation and urban migration. The initiative to set up a professional team was announced by Ōita governor Hiramatsu Morihiko to the local assembly in 1994, making the objective of becoming a World Cup host city explicit. Since the World Cup was coming to town, the very active prefecture government used the opportunity to channel large amounts of public funds into expanding the network of motorways and railway tracks connecting the basically rural city with the busier cities in Kyushu's northwest and the main island of Honshu. In relative terms, costs for the Big Eye, Ōita's new World Cup stadium, were considerably low. Following the national government's lead, the greatest part of the total funds to cover the construction costs was collected from general obligation bonds issued by Ōita Prefecture.

Like Kashima, football was welcomed as a solution to the problems of unbalanced regional development. However, the region possessed neither a semi-professional company team nor the financial support of a potent sponsor company. In comparison, neighbouring Fukuoka had more attractive conditions to offer and hence was able to convince an entire team (Avispa Fukuoka's amateur predecessor) from Shizuoka, where some attractive J-League teams such as Shimizu S-Pulse and Jubilo Iwata already were based, to move to western Japan. The Ōita team, however, had to be built up from scratch, largely using local talent from high school and university teams. Support was mandatory for all the prefectural government public employees and they were automatically added to the team's fan club. Public sponsorship for the football club thus could take the form of buying tickets en bloc for the supporter group. Government employees were also sent as temporary staff workers to run the club office and manage team affairs. A career track bureaucrat from the Ministry of Local Autonomy in Tokyo was officially dispatched as General Manager to help establish Ōita Trinity. Since the home region lacked any large corporations, the bureaucrats solicited sponsorship fees from numerous small and medium sized companies in the region and even succeeded in recruiting sponsors from the main island (Kimura 2003). This resulted in Ōita Trinity's successful transformation from a local amateur club into a professional team. Ōita's football community took more time to create, even though Trinita, the slightly changed name of the professional team, had been selected to appeal to a trinity of sponsors from the business world, local administration and citizens in the prefectural region. According to an Ōita City leaflet: 'Oita Trinita have total prefectural support, since not only the prefecture of Oita but also enterprises, companies, and all the inhabitants support it.' Research on the fan communities in the city suggests that grassroots football fan groups neither appreciated local elites' promotional role in running the official supporters club, nor the involvement of the bureaucracy in management affairs (Yamashita and Saka 2002). However, with the new World Cup stadium in town and the prospect of promotion to J1 ahead, Ōita turned out to be one of the best supported J2 teams in 2001 and again in the following year when it was promoted to J1.

The most supported team in J2 in 2002, and according to a recent fan survey carried out by the J-League, the one with the oldest supporters, hails from Japan's northeast, Niigata (J-League 2006). Professional football's development there is another example of community formation through football. No other place in Japan is more closely associated with political corruption and pork barrel politics than Niigata, a remote industrial city and the homonymous capital of the prefecture on the coast of the Japan Sea. The Joetsu Shinkansen track, for example, is a relic of the former premier Tanaka Kakuei who channelled large amounts of public funds into underdeveloped regions. Despite huge construction projects, Niigata continued to be regarded as a rural, backward (*inaka*) and boring (*tsumaranai*) location, with *koshihikari* rice providing the only source of local pride (Uchiumi 2004: 121). Discussions about forming a professional team started in Niigata as early as 1994, since the local authorities nurtured a strong desire to possibly host some World Cup games. In contrast to Ōita, Niigata's initial impetus came from the local business elite. Ikeda Hiromu, head priest of a local shrine who had made a fortune as the owner of a network of prep schools, became the first president as well as the new team's main sponsor.

The local club's history goes back to the Niigata Eleven Soccer Club – founded in the mid-1950s and later advancing to the regional amateur league – which merged with some amateur teams from the region. Supported by thirty influential companies in the prefecture and funds provided by around 150 local companies and organizations, ¥500 million were raised to begin the move toward full professionalism. As with Ōita Trinita, Niigata Albirex joined the J-League as an inaugural member of the newly established J2 division. After three unremarkable seasons Niigata's Big Swan World Cup Stadium opened in 2001. Once Albirex moved its home games into the Big Swan arena, it became one of the best-supported teams in the J-League. Regular attendance at home games of more than 35,000 exceeds the average for most other J1 and J2 clubs by a long way. In 2002 Albirex faded at the final stretch, but in 2003 it captured the league title and gained promotion to J1. Support did not weaken thereafter, although it has been common to see a sudden decline once the race to the top has been won, and the newcomer struggles to hold pace with the established J1 teams.

According to Kōzu's research team (2002), the club's management was extremely successful in raising local support because it placed great emphasis on local culture and communication. More than in any other region, Niigata was promoted and received as the locals' team, even though in terms of percentages Niigata is closer to the average (male supporters 52.9 per cent, average age 36.6 per cent, according to the J-League Supporters' Survey 2005) of young and male-based football support than Ōita, for example. Yet there are simply more fans than in any other stadium, inviting television documentaries, match reports and popular accounts to emphasize the local roots of football support in Niigata and football's role in community building. However it is an open secret that a fully packed stadium does not necessarily mean riches for the team. The club has become notorious for giving away free tickets to locals. Initially tickets were distributed without a clear strategy; then the club's management began to make effective use of neighbourhood associations' contacts and traditional networks. For example, as soon as the team entered the professional league, Albirex set up a dense network of supporter or booster clubs. Niigata prefecture was divided into more than forty sections, and each section saw the opening of its own fan club. In many instances, those who became the directors were usually also board members of other local associations, and these networks were used for spreading the word about Niigata's football team as well as for the distribution of free tickets. Similarly, information on match days and available tickets was passed on through the traditional Japanese *kairanban*, an information bulletin that circulates from household to household to inform residents about important news (Tsujiya 2005: 170). Giving free tickets away to young children looked like a thinly disguised attempt to lure their paying parents or grandparents into the stadium, but the club management reasonably expected to raise awareness of the local team once people from Niigata had actually experienced the impressive atmosphere in the Big Swan stadium.

Despite its traditional and parochial emphasis, Niigata's football community is not exclusive. It is inclusive because it allows different people from various backgrounds – bank executives, housewives, school children, volunteers, long-time residents and newcomers – to interact with each other. This aspect is heralded as a distinctive feature by the J-League since it establishes communal bonds solely on the basis of residence. It is not a question of provenance, of regional ancestry – another traditional variable in Japan's community building – since we have talked to and met numerous Albirex fans within the Niigata region who have moved in only recently. This is a novelty since Japanese hometowns were seen to be exclusive, its inhabitants embodying values that were not easily communicated. As a social project, shopping was found to be among the five most frequent activities Niigata's team has also found favour with various artistic groups, including the popular rock band, The Penpals, who remixed their national smash hit song 'Believe' to capture the attitude of the Niigata supporters. They recorded a live performance with a 30,000-plus background choir, dressed in Albirex's team colour, orange, expressing their shared belief in, and love for, the team (Morisaki 2004).

Niigata Albirex now has been adopted as a new symbol for the entire prefecture, although under very unfortunate circumstances. In October 2003, during Albirex's first year in the J1, an earthquake devastated its hometown region. In the Niigata Chuetsu Earthquake's aftermath, all

J-League clubs, players and supporters contributed to disaster relief efforts. Football fans donated generously at every stadium, including the collection of ¥4 million at a single Urawa Red Diamonds home game. Moreover Albirex players and staff members travelled throughout the region in order to console and entertain displaced families. They gave generously to the victims, contributed to a charity auction organized by the players' association, and played an exhibition match against a Dream Team of past, present and future national team players. In return, fans from the entire Niigata prefecture, which had been recently identified as the team's new official hometown region, have continued flocking to the Big Swan (J-League 2005a). It is this kind of mobilization and reaction to a local disaster that illustrates the position of some football clubs in this phase of the J-League's development.

## The Making of Football Communities in Japan: Phase 3

Since the 1999 launch of the J2, twelve new clubs have been promoted from the amateur Japan Football League (JFL) into it, and the J1 has expanded from sixteen to eighteen teams. Discussions were held about the possibility of introducing a professional third division on a regional level and it was decided in February, 2013, to launch this third division (J3) in 2014 'for the further expansion of the number of J-League clubs to serve as catalysts for the development of sporting culture' (J-League 2013: 7).[5] Suzuki Masaru, the former J-League chairman, expressed the wish to build up to a hundred clubs, or approximately two in each of the forty-seven prefectures

5 J3 will consist of clubs, which satisfy certain inspection conditions, with J. League affiliate membership and they will compete as with the aim of gaining promotion to J2. Affiliate members aiming to join J3 were called J-League 100 Year Vision Clubs, and the J-League will support these clubs in their efforts to qualify for J3 membership (J-League 2013: 7–8).

throughout Japan. While the corporate consortium approach is still widely practised, the continued Japanese fiscal crisis has forced local authorities to save rather than spend, and the failed attempts to develop clubs from Kagoshima (Volca) and Okinawa (Kariyushi FC) have shown that sustainable alternatives to a reliance on sponsorship income are essential for newcomers. Three of the more recent new entrants – Mito Hollyhock, Yokohama FC and Thespa Kusatsu – illustrate alternative ways for the establishment of professional football clubs.

## Community Building: From the Bottom Up?

Mito Hollyhock introduced a *socio* (partner) style of membership system – a one member one vote arrangement similar to FC Barcelona – which helped the club to get approval for the J-League. Links with established J-League clubs, including Yokohama Marinos and FC Tokyo who have farmed out young players to Mito as a way of helping them get accustomed to life in the J-League, have kept start-up expenses low, while financial subsidies from Ibaraki prefecture saved the club when it faced its most severe financial problems, to date, in 2004. Yokohama FC emerged as a direct result of the decision to dissolve AS Flugels at the end of the 1998 season. It is therefore the first fan-initiated football club in Japan that also started with a socio-style system.[6] With increasing success, sponsorship money came flooding in and reduced the reliance on the socio members. In recent years however a gap has opened up between club management and the socio members who are demanding more involvement in club affairs than the management has so far been willing to concede.

Thespa Kusatsu was launched in 1997 in the small town of Kusatsu in Gunma prefecture, popular for its *onsen* (hot spring) and with fewer

6 Similar in some respects to FC United, the team formed by disgruntled fans following the takeover of Manchester United by Malcolm Glazer in 2005.

than 10,000 inhabitants. While the small population size itself is a remarkable breach of (former) J-League regulations, the business model is even more interesting. Until the team was promoted to play in the J2 in 2005, most players were working part-time in local shops and hotels, but not for their own income. In the case of Thespa, all sponsorship money came from the local economy, which in return received the athletes' labour. Thespa management thus used the sponsorship income to pay players' salaries and its operational expenses. Kusatsu town supported the team by providing free access to the local sport facilities and by subcontracting the municipal Beltz Hot Spring Centre's management team to the club (Tsujiya 2005: 97, 124). Since its promotion Thespa has moved out of tiny Kusatsu, not least because of the J2 requirements on stadium size. The closest appropriate stadium, with a capacity of 15,000, could only be found over two hours' ride away by car in Maebashi, Gunma prefecture's capital city. Hence Thespa Kusatsu has come to represent the towns and cities of Shibukawa, Takasaki and Maebashi, as well as Kusatsu, which continued to serve as the home base of Thespa's satellite team for some years. Meanwhile, the club was renamed Thespa Kusatsu Gunma in 2013. Extending the hometown area is not unknown. In the case of JEF United (which arguably holds the dubious record of garnering the least spectator interest in its fourteen year existence) the hometown area of Ishihara City was officially extended in 2004 to include neighbouring Chiba, capital of the eponymous prefecture, which is now officially part of the team's name and also provided a brand new 18,000 person capacity stadium. Kashima Antlers has expanded its hometown area since 2005 as well, for the purpose of raising interest among local inhabitants and potential investors. In 1998, Bellmare Hiratsuka relaunched itself under the name Shonan Bellmare, a larger geographical area along the coast of Kanagawa Prefecture, after its main sponsor Fujita withdrew (Kawabata 2003: 60). Municipal mergers caused by administrative reforms have also affected clubs' allocation. In 2003, Shimizu City became a district of the much larger Shizuoka City, the current home town of Shimizu S-Pulse. Urawa city was combined with Omiya and Yono to form Saitama City in 2001, which is now home to two J-League teams – the Urawa Reds and Omiya Ardija.

Such transfers, removals and name changes have sometimes affronted supporters and fans, and annoyed the local authorities and sponsors waiting for football clubs' promised contribution to local social amenities and communities. A study conducted in early 2003 among seven J-League localities in western Japan also revealed that most of the announcements in that region regarding football-community relations had up to then consisted of little more than lip service to the J-League mission statement, largely because of many football teams' constrained financial condition (Manzenreiter 2004b). It was around this time that the J-League intensified its efforts to promote the ideal of comprehensive communal sports clubs. Clubs were invited to apply for J-League support for programmes aiming at the expansion of their wider sport programme. In 2002, for example, Shonan Bellmare established a separate, non-profit sporting organization with a focus on triathlon, beach volleyball and football for under 15-year-old players to foster bonds with the local community and to defray the risks associated with sport entertainment and sport social education (Kawabata 2003). Albirex Niigata supports a relay marathon (*ekiden*) team, a basketball team, cheerleaders and a winter sports club. Urawa Reds became the first J. League club to open a comprehensive centre for community sport, Redsland, which is managed by the club and has five football pitches, three of which can also be used for baseball, a rugby pitch, eleven tennis courts, a daytime camping area and a farming zone. 'Urawa hope that this composite park will foster sporting activity among the young, encourage lifelong participation in sport and serve as a focal point for community development' (J-League 2005b: 5). Futsal (a form of five aside football) pitches and other facilities have been added, since full-scale operation began in 2007. Through this J-League sponsored programme, some teams have turned into multi-sport organizations, although it is difficult to say exactly whether this kind of expansion significantly triggered mass sport participation.

As the J-League seeks to instil its ideologies and visions for football-community relations at the grassroots in Japan the main efforts have been concentrated on further promoting football amongst children and young people. J-League membership contracts were changed so that community involvement by all registered clubs and players has become mandatory since 2003. Since then, J-League players, coaches and other club staff have joined in thousands of community activities that include soccer schools, visits to children's and old people's homes, and other forms of social service. Annually held 'JOIN Days' have provided special family events, including football training, during the national holiday period at the end of April and beginning of May (J-League 2004: 7). Literally at grassroots level, the J-League has also launched a campaign to increase the number of grass pitches available in elementary and high schools. 'Mr Pitch', a two metre tall official mascot – a walking piece of (artificial) grass – has become the 'J-League 100 year Vision Messenger' that features prominently on the J-League website, greeting cards and other promotional media. As the campaign's icon, Mr Pitch has been a regular visitor to events organized by the J-League and its member clubs (see <http://www.j-league.or.jp/100year>).

## Conclusions

If belonging is 'about participation in communication' (Delanty 2003:187), then the J-League has started something. There appear to be many Japanese people who aspire to belong to the imaginary football family, despite our emphasis here, on top-down rather than on bottom-up developments in football – and we think that this is a fair reflection of a Japanese culture of 'friendly authoritarianism' (Sugimoto 1997). Hence the creation of football communities is firmly linked to a tradition of initiatives from the bottom up, while top down initiatives are more widely institutionalized and socially accepted. We hope to have indicated ways in which the imagined football community can be both constitutive of agency as well as experienced as an imposed institutional structure. This is evidenced by the increasing number of players, officials and teams formally registered with the JFA. In 2012 nearly 30,000 11-a-side teams have been registered, and there were over 1.4 million players and officials – including 953,740 players and 125,436 futsal players (see <http://www.jfa.or.jp/jfa/databox/index.html>) – who enjoy a less formalized way of practising the game, a massive increase beginning from the 2002 World Cup. The initiatives discussed in this chapter reveal how football in Japan is working to give something back to society. Having identified a number of key variables responsible for the emergence of distinct patterns of community-football relations in Japan, we have shown that whilst much of the discourse on football and the Japanese community serves very different interests, it continues to facilitate most of all the proliferation of private consumption, public spending and transnational money flows. Discussing community development and hinting at community regeneration and improvement through community mobilization around football, are just two of the methods being used to create public support and to gain wider acceptance for the J-League mission.

Any successful creation of a sports community relies upon mediated coverage, and Japan's commercial mass media have been prominent allies, patrons and sponsors of certain sports (Horne 2005). However football still struggles against the prominence of other sports shown on Japan's broadcasting networks. Football in Japan has attracted a loyal fanbase in the past twenty years, but its media coverage is still largely dictated by commercially driven estimates about the audience ratings compared with those for baseball and sumo (Wong et al. 2013). In this respect it is the new media – internet and mobile telephony especially – that have established networks amongst clubs, supporters and peripheral visitors. Interactive media particularly enabled the creation of notable cyber-fan communities, following football as well as other spheres that the grassroots football community will most likely develop in the immediate future.

In this chapter we have argued that the J-League has been successful both in setting up a prospering industry and in shifting the collective Japanese imagination away from the nation to the region and the locality while at the same time managing to involve the nation in the global love affair with football. Since the 1990s Japan has joined the rest of the world in celebrating *the* global sport, and in doing so it has also encountered football's distinctive relationship with the mass media and other transnational networks that has led the sport to become more important and financially lucrative (Cleland 2015). Making professional football a viable business in Japan has been a challenge, and for many clubs will continue to be so. As a product, and a consumer good, however, football has clearly come to stay. As the football journalist Keir Radnedge (2005b: 6) noted: 'Japanese football and the J-League, both through the teams and the individual players, have come of age in European eyes.'

## References

Arimoto, T. (2004). 'Narrating Football'. *Inter-Asia Cultural Studies*, 5(1): 63–76.

Bale, J. (2000). 'The Changing Face of Football: Stadiums and Communities'. In J. Garland, D. Malcolm and M. Rowe (eds), *The Future of Football*, pp. 91–101. London: Frank Cass.

Bestor, T. (1989). *Neighborhood Tokyo*. Tokyo: Kodansha International.

Castells, M. (1996). *The Rise of the Network Society*. Oxford: Blackwell.

Cleland, J. (2015). *A Sociology of Football in Global Context*. London: Routledge.

Delanty, G. (2003). *Community*. London: Routledge.

David Taylor Partnerships (DTP). (2006). *Active Engagement: A Study of Northwest Professional Sport Clubs' Involvement in Community Regeneration*. Warrington: NWDA.

Football Task Force. (1999). *Investing in the Community. A Submission by the Football Task Force to the Minister of Sport*. London: Football Task Force.

Fukutake, T. (1989). *The Japanese Social Structure – its Evolution in the Modern Century*. Toyko: University of Tokyo Press.

Harvey, D. (1989). *The Conditions of Postmodernity*. Oxford: Basil Blackwell.

—— (1997). 'The New Urbanism and the Communitarian Trap'. *Harvard Design Magazine* 1:1–3.

Hirose, I. (2004). 'The Making of a Professional Football League: the Design of the J-League System'. In W. Manzenreiter and J. Horne (eds), *Football goes East: Business, Culture and the People's Game in East Asia*, pp. 38–53. London: Routledge.

Horne, J. (1996) '*Sakka* in Japan'. *Media, Culture and Society* 18(4): 527–47.

—— (1999). 'Soccer in Japan: Is *Wa* all You Need?' *Culture, Sport, Society* 2(3): 212–29.

—— (2001). 'Professional Football in Japan'. In J. Hendry and M. Raveri (eds), *Japan at Play*, pp. 199–213. London: Routledge.

—— (2005). 'Sport and the Mass Media in Japan'. *Sociology of Sport Journal* 22(4): 415–32.

—— with D. Bleakley (2002). 'The Development of Football in Japan'. In J. Horne and W. Manzenreiter (eds), *Japan, Korea and the 2002 World Cup*, pp. 89–105. London: Routledge.

—— and D. Jary (1994). 'Japan and the World Cup – Asia's First World Cup Finals Hosts?'. In J. Sugden and A. Tomlinson (eds) *Hosts and Champions: Soccer Cultures, National Identities and the World Cup*, pp. 151–68. Aldershot: Avebury Press.

Horne, J. and W. Manzenreiter (2008). 'Football, *Komyuniti* and the Japanese Ideological Soccer Apparatus'. *Soccer & Society* 9(3): 359–76.

J-League. (2004). *J-League News: Official Newsletter* 30. Tokyo: Japan Professional Football League.

—— (2005a). *J-League News: Official Newsletter 31*. Tokyo: Japan Professional Football League.

—— (2005b). *J-League News: Official Newsletter* 33. Tokyo: Japan Professional Football League.

—— (2006). *2005 J-League Sutajiamu Kansensha Chōsa Hōkokusho*. Tokyo: Nihon Puro Sakkā Riigu.

—— (2013). *2013 J-League Profile*. Tokyo: J-League.

Kawabata, Y. (2003). '*Sekkeizu o Tsuita Yume. NPO Hōjin Shonan Bellmare Supōtsu Kurabu*'. *Sakkā Hihyō* 17: 59–67.

Kimura, Y. (2003). '*Oita Trinita, Kanshudō no Ketsujitsu to Kongo*'. *Sakkā Hihyō* 18: 70–5.

Koiwai, Z. (1994). '*Sakkā ni Yoru Machizukuri*'. *Toshi Mondai* 85(12): 59–69.

Kōzu, M. et al. (2002). *Puro Supōtsu to Chiiki Chakumitsu. Supōtsu Chiiki Chōsa in Niigata*. Tokyo: Hitotsubashi University <http://www.soc.hit-u.ac.jp/~kozu/activity/> accessed 30 March 2006.

Kubotani, O. (1994). '*Supōtsu shinkō ni yoru kiban seibi. Genjō to kadai*'. *Toshi Mondai* 85(12): 43–57.

Lützeler, R. and E. Ben-Ari (2005). 'Urban Society'. In J. Kreiner, U. Möhwald and H. D. Ölschleger (eds) *Modern Japanese Society*, pp. 277–303. Leiden: Brill.

Manzenreiter, W. (2002). 'Japan und der Fußball im Zeitalter der technischen Reproduzierbarkeit: Die J-League zwischen Lokalpolitik und Globalkultur'. In M. Fanizadeh, G. Hödl and W. Manzenreiter (eds), *Global Players. Kultur, Ökonomie und Politik des Fußballs*, pp. 133–58. Frankfurt: Brandes & Apsel/Südwind.

—— (2004a). '*Nihon Shakai no 'Ibentoka' to Sakkā*'. *Supōtsu Shakaigaku Kenkyū / Japan Journal of Sport Sociology* 12: 25–35.
—— (2004b). 'Sport zwischen Markt und öffentlicher Dienstleistung. Zur Zukunft des Breitensports in Japan'. *SWS-Rundschau* (*Journal für Sozialforschung*, Special Issue) 44(2): 227–51.
—— (2006). 'Fußball und die Krise der Männlichkeit in Japan'. In E. Kreisky and G. Spitaler (eds), *Fußball: Die männliche Weltordnung*. Frankfurt: Campus.
Manzenreiter, W. and J. Horne (2002). 'Global Governance in World Sport and the 2002 World Cup Korea/Japan'. In John Horne and Wolfram Manzenreiter (eds.) *Japan, Korea and the 2002 World Cup*, pp. 1–25. London: Routledge.
—— (2005). 'Public Policy, Sports Investments and Regional Development Initiatives in Contemporary Japan'. In J. Nauright and K. Schimmel (eds), *The Political Economy of Sport*, pp. 152–82. London: Palgrave Macmillan.
—— (2007) 'Playing the Post-Fordist Game in/to the Far East: the Footballisation of China, Japan and South Korea'. *Soccer and Society* 8(4), 561–77.
Millward, P. (2011). *The Global Football League*. London: Palgrave Macmillan.
Morisaki, K. (2004). '*Believe – Keiyaku no Shunkan*'. In S. Hihyō (ed.), *Niigata Genshō. Nihonkai Tenkoku no Tanjō o Megutte*, pp. 195–235. Tokyo: Futabasha.
Robertson, J. (1991). *Native and Newcomer: Making and Remaking a Japanese City*. Berkeley: University of California Press.
Sawano, M. (2005). *Kigyō supōtsu no Eikō to Zasetsu*. Tokyo: Seikyusha.
Sugimoto, Y. (1997). *An Introduction to Japanese Society*. Cambridge: Cambridge University Press.
Tsujiya, A. (2005). *Sakkā ga Yatte kita. Thespa Kusatsu to Iu Jikken*. Tokyo: NHK Shuppan.
Ubukata, Y. (1994). *J. Riigu no Keizaigaku*. Tokyo: Asahi Shinbunsha.
Uchiumi, K. (2004). '*Arubi ga Seichō o Tsuzukeru Tame ni*'. In S. Hihyō (ed.), *Niigata Genshō. Nihonkai Tenkoku no Tanjō o Megutte*, pp. 111–28. Tokyo: Futabasha.
Watts, J. (1998). 'Soccer *Shinhatsubai*: What are Japanese Consumers Making of the J League'. In D. P. Martinez (ed.), *The Worlds of Japanese Popular Culture*, pp. 181–201. Cambridge: Cambridge University Press.
White, M. (2002). *Perfectly Japanese: Making Families in an Era of Upheaval*. Berkeley: University of California Press.
Wong, D., I. Kuroda and J. Horne (2013). 'Sport, Broadcasting and Cultural Citizenship in Japan'. In J. Scherer and D. Rowe (eds) *Sport, Public Broadcasting and Cultural Citizenship: Signal Lost?*, pp. 243–62. London: Routledge.
Yamashita, T. and N. Saka (2002). 'Another Kick Off: the 2002 World Cup and Soccer Voluntary Groups as a New Social Movement'. In J. Horne and W. Manzenreiter (eds), *Japan, Korea and the 2002 World Cup*, pp. 147–61. London: Routledge.

GRISELDIS KIRSCH

# Relocating Japan? Japan, China and the West in Japanese Television Dramas

## The Asia Boom during the 1990s

Japan 're-discovered' its Asian neighbours when the world order collapsed at the end of the Cold War in the late 1980s.[1] After a long era of looking towards the West, which occasionally had culminated in admiration (Hijiya-Kirschnereit 1988; Creighton 1995, 1997), other Asian countries suddenly gained importance for Japan. What previously had been only an economic relationship evolved into a virtual Asia boom in the cultural sphere. Concurrently, more Asian immigrants entered Japan,[2] so that Asia became more visible and Japan's 'myth' of a homogenous society began to be challenged even more than ever (Weiner 1997).

These new societal changes soon were represented in the media. Alongside a growing number of news features on the alleged increase in Japan's crime rate – a result, it was claimed, of more crimes committed by foreigners (*gaikokujin hanzai*, see Yamamoto 2004 and Herbert 2002) –

1 In my *Contemporary Sino-Japanese relations on Screen. A History, 1989–2005* (2015) I look at the historical development of the representation patterns of Greater China throughout the period of China's economic growth. Here, I draw on examples that I also analyse in the book, however here the focus is set on Japan's imagined identity between Asia and the West. This paper was originally prepared before the release of the book. The argument in the book follows the patterns of historical representation – while the focus of this paper is on the examples as such.

2 See Statistical Yearbook of Japan 2008, <http://www.stat.go.jp/english/data/nenkan/1431-02.htm>.

and documentaries on life in Asia (Gatzen 2002), fictional media such as cinema and TV drama also responded to this and picked up more Asian topics. Among the fictional genres, Japanese cinema was the first to embark on this trend and a large number of films featuring Asian characters were produced. As Schilling notes '[t]hroughout the 1990s, one of the topics [young Japanese filmmakers] have found most fruitful is the growing intercourse between Japan and its Asian neighbors' (1999: 41).

By 2000 television dramas, the most important fictional genre on the small screen, reacted to these developments. Until the 1990s, Japanese television dramas had focused on domestic issues such as family problems and social change (Gössmann 2000), but its themes began to shift gradually and transculturality became a new trend (Gössmann and Gatzen 2003; Kirsch 2007, 2015). Thus, the number of foreign characters, among them a large number of Asians,[3] first in television films and later in television series began to increase significantly. While until 1994 the share of dramas with foreign characters amounted to less than 1 per cent, it suddenly rose well above this mark,[4] indicating an apparent shift towards a more 'international' small screen. Although the audiences for TV dramas have been declining recently in Japan,[5] the genre remains vitally important for understanding the Asia boom since it is said to be especially close to society, acting, in

3 In this context, Asia or Asians refers predominantly to people from other East or South (East) Asian countries – it does not refer to the geographical designation.

4 These numbers are based on my own statistical analysis of *Terebi dorama zenshi 1954–1994* (Tokyo News Mook 1994), which lists every single drama aired during this period. I also used *Rendorama 10 nenshi* (*Za terebishon* 2004) as another point of reference. Although the latter work provides only a representative average rather than a complete overview, it demonstrates that after the turn of the millennium, the number of dramas set in Asia or with an Asian character has risen significantly, while remaining far below that of dramas focussing on Japanese topics. However, this Asia boom was first located within a more international boom that saw an increase in dramas with foreign characters in general.

5 In his overview of the history of Japanese television drama, Okada tells the genre to 'hold on' and get out of the crisis (2005: 169).

the words of the media researcher Rogge, as a 'psychometer of societal and individual circumstances' (1987: 74, my translation).[6]

Moreover in 2003 the public broadcasting station NHK aired the Korean TV drama *Fuyu no sonata* (Winter Sonata) to great success, leading to the unprecedented popularity of Korean stars in Japan. As the drama's protagonist, Yong-joon Bae quickly became a star in Japan.[7] Since then, many Korean stars have appeared in Japanese TV dramas and Korean dramas have been remade as Japanese productions; while Korean pop music has also become more visible, and audible, on television. Therefore, a medium which has been focused on Japan for more than thirty years has recently become more internationalized.

These more recent dramas are not the topic of this paper; rather I will examine the dramas developed at the beginning of the Asia boom, which paved the way for these latter developments. With these earlier dramas – though not all were necessarily commercially successful – the producers of television dramas took their first steps towards a more international genre. At that time it was not Korea, but the regions of Greater China[8] that were portrayed most often and Chinese actors and actresses appeared on the small screen. For that reason, it is illuminating to look at these dramas and see how the characters' Otherness (i.e. Chineseness) was appropriated within the plot.

The following analysis will reveal that these dramas were not solely about Japan and China, but also implicate the West in their stories. In a way, these representations negotiated Japan's post-war identity which 'has [...] always been imagined in an asymmetrical totalizing triad between "Asia", "the West", and "Japan"' (Iwabuchi 2002: 7). Thus, we might ask:

6 The *Daily Yomiuri* columnist William Penn (2004) argues that they provide better insight into Japanese society than any sociology textbook.

7 On the Korea boom see Hayashi 2005, Hayashi and Lee 2007 as well as Gössmann 2008.

8 The term China or Greater China is used to denominate characters from the People's Republic of China, Taiwan and Hong Kong. I will analyse one drama with one character from each of the three regions. The dramas have been chosen because of their representativeness in this context.

how is Japan situated within this triad? Is Japan a western country, an Asian country, something on its own or a hybrid?

Identities, whether of the Self or the Other, need to be continually constructed (Bauman 1997) and imagined communities (Anderson 1994), such as that of any modern nation state, need a constant (re-)narration of their myths and identities (Bhabha 1990, 2007). Since TV drama can be seen as 'a primary generator and the most everyday source of narratives' (Thornham and Purvis 2005: ix), the media, particularly fictional genres, due to their descriptive character, are especially important since they provide a perfect vehicle for these narrations (cf. Martinez 1998). Thus, it is important to first look at the relation between Japan and the West as well as Japan and Asia, since it can be assumed that it also might be mirrored in the media.

## Admiration and Delimitation: Japan and the West

*Nihonjinron* (theories of the Japanese) authors have tried to delineate the supposed uniqueness of Japanese culture and the Japanese (Befu 1992, 2001; Miller 1982), constructing Japan as a homogeneous country. *Nihonjinron* literature has tended to contrast Japan with the West elaborating on the dissimilarities between the two cultures. A tendency to homogenize the West is also typical, so that an academically profound discussion on different cultures is not to be expected. Moreover, as Befu (2001) argues, the West has long looked upon Japan as a monolithic entity and *nihonjinron* were merely a response.

Thus *Nihonjinron* do not clearly map out the cultural boundaries within the West; rather they are ignored in favour of the negotiation of difference between the West and Japan. The most significant dissimilarity between Japan and the West is alleged to be Japan's collectivism and groupism in contrast to the West's individualism (Gudykunst and Nishida 1994: 20; Befu 2001). A common argument is that whereas the Japanese will always work for the benefit of the group (with boundaries that encompass nation, family, company), Westerners will always protect their own individual interests rather than that of the greater good. In tandem with this depiction, there also exists a sense that (post)modernity is a western predicament; that is, it leads to a loss of values and creates social problems – whereas Japan is said to have kept to its (Confucian) traditions despite having become an industrialized, modern nation. Given that these theories also are found in western publications on Japan, the question arises as to which is the chicken and which is the egg (Revell 1997).

Controversial or not, oversimplified or not, over the course of the time, the ideas promulgated in *nihonjinron* literature have become key in the formation of Japanese identity. The concept of Japanese uniqueness even led Huntington (1993), in his work on the clash of civilizations, to argue that Japan is the only nation-state that is also a civilization, thereby granting it a very exceptional status in the world (Antoni 1996). While admittedly *nihonjinron* as a genre has to some extent lost relevance in the explanation of 'Japan', the stereotypes perpetuated therein are still highly relevant.

This is only one side of the story. Given the long relationship between Japan and the United States, there has also been a distinct admiration for the United States (which is often seen as *pars pro toto* for all western countries). Post-war, the influence of the United States has been considerable; and as their relationship, albeit asymmetrical in power, has grown more intense, the catchphrase internationalization (*kokusaika*) has gained currency. Consequently Japan was supposed to open further to the West. While the Japanese economy could already be labelled international because of its global commitments, additional segments of Japanese society were also expected to become more international (cf. Hook and Weiner 1992). Critics such as Katō (1992) have argued, however, that the process was never one of true internationalization, but of westernization and the (white) West became *the* model to emulate (see also Russell 1994). Furthermore, Katō (ibid.) notes that over time an international person (*kokusaijin*) came simply to be seen as somebody who spoke English. Underlining the importance of the United States within this alleged internationalization were the Hollywood stars who made and still make frequent appearances in television commercials (Creighton 1995, 1997; Kirsch 2001; Prieler 2006).

Moreover western models were seen as the epitome of beauty, while western experts explained to Japanese audiences how to apply beauty products. In these representations a hierarchy in which Japan is placed below the West is clearly constructed; Japan is shown as still learning and struggling to keep up with it. Recently however a tendency to ridicule the West has begun to be observed in Japanese television advertising.

The relationship between Japan and the West could be termed ambiguous: clearly dominated by the need to delimitate one's own culture as expressed in the *nihonjinron* on the one hand and the admiration for things western on the other. An encounter with the West on an equal footing has yet to take place. In contrast, the relations between Japan and its Asian neighbours seem similarly problematic at first glance, yet the reasons for this are arguably different.

## Delimitation but no Admiration? Japan and its Asian Neighbours

During Japan's post-war history, its relations with Asia have been strained, to say the very least. Japan's past as the aggressor in Asia, exercising a brutal colonial rule over its neighbours before the war and then later post-war governments' failure to apologize for the war crimes committed by the Japanese Army in the countries it had invaded, has put a strain on its inter-Asian relationships. Japanese prime ministers' visits to Tokyo's Yasukuni Shrine, where war criminals are enshrined next to ordinary soldiers, as well as the publication of textbooks glossing over Japan's wartime aggressions, have frequently led to anti-Japanese protests in other Asian countries, most notably in the People's Republic of China and Korea (Wieczorek 2001; Richter 2004). Despite this volatile political relationship, economic ties are increasing in complexity.

Recently, despite the political problems and anti-Japanese protests, China has become one of Japan's most important trade partners and also one of its most popular tourist destinations.[9] Therefore, although China has not played a major part in Japanese foreign politics for much of its post-war history, due to the fact that the People's Republic of China belonged to a different political bloc, it now increasingly is doing so.

A certain ambivalence in the representations of Asia in Japanese popular culture is evident. On the one hand, there is the portrayal of the threat posed by Asia's alleged anti-Japanese-ness as well as the mediation of the heated debate on crimes committed by Asian foreigners in Japan; in both cases it is notably the Chinese who are depicted as slowly undermining the Japanese state, erecting quasi-exterritorial spheres.[10] On the other hand Asia and all its constituent countries are seen as nostalgically embodying 'a social vigor and optimism for the future that Japan allegedly is losing or has lost' (Iwabuchi 2002: 159). Not insignificantly, the latter notions appeared as the Japanese economy stumbled into crisis in 1991– while the economy in Asia began to boom. Hence, the will to modernize which allegedly has been lacking in Japan since the recession, is now assumed to be found in the rest of Asia.

This can also be seen in the representations of Asia in the Japanese media. As Gatzen and Gössmann (2003) have noted, in both television dramas and documentaries, Asia tends to be regarded as having retained a certain kind of energy. It would appear that representations of other Asian countries oscillate between visions of horror (anti-Japanese protests and

9 In July 2007, 308,000 individuals travelled to the United States, while 327,000 went to the People's Republic of China. Approximately one year later, the number of tourists to the People's Republic of China had declined considerably, while Macau and Malaysia saw more Japanese tourists than ever. A certain shift of travel destinations within Asia seems to be occurring. See <http://www.tourism.jp/english/statistics/outbound.phpt>.

10 In February 2006, a *mook* (Japanese-English for a single-issue magazine) on foreign criminality appeared, leading to immediate protests from foreigners in Japan that included a boycott of stores selling the *mook*. In this magazine, all foreigners are awarded a degree of dangerousness (*kikendo*), in which the mainland Chinese rank first. The *mook* also elaborated on the issues of areas turning into exterritorial spaces (see Washinton 2006; Kirsch 2015).

*gaikokujin hanzai*) and a longing for their spiritual richness (Kirsch 2007 and 2015).

Comparing the, albeit short, outline of the relations between Japan, Asia and the West, it becomes obvious that none of the two pivots in the 'totalizing triad' seem to have a sunny relationship with Japan. Thus, either side is appropriated in a Lacanian sense in order to construct the Japanese Self; Otherness is used as a foreign mirror in order to construct a Japanese identity. The West is used as a model to emulate or deviate from, whereas Asia becomes both the domain of things lost and the source of a threat to Japan. In this sense, Japan's images of the Other seem to oscillate between what the German sociologist Hahn (1992; 1994) terms *fascinosum* and *tremendum* (fascination and horror). This representation also corresponds to Faulstich's (1996) analysis of the three patterns in the construction of Otherness in film: exoticism, salvation and horror. Here, particularly the horror and salvation patterns seem to be applicable.[11]

Analysing the representations of Asia, Japan and the West in fictional genres through their narrative structures we see that they are used to construct Otherness in exactly this dual manner: as a role model, or as a threat.[12] The following analyses aim to delineate how Japan is portrayed in relation to Asia and the West and what role westernization plays in the examples. Since the regions of Greater China were of vital importance in the Asia boom, I have chosen China as focal point, starting with one of the first dramas aired during the Asia boom on television, the 2001 series *Uso koi* (False Love).

11 These patterns also correspond to Said's (1978) concepts of Orientalism. Like the West, Japan seems to orientalize the Other.

12 Horror is of less relevance in these dramas. See Kirsch 2015 for an overview of the relevance of the horror pattern.

## A Westernized Chinese in Japan: *Uso koi*

*Uso koi* was broadcast by the private television station Fuji TV. The actors Kiichi Nakai and Yukie Nakama starred and the (Hong Kong) Chinese star Faye Wong made her first appearance on the Japanese small screen (see also Gatzen and Gössmann 2003; Gössmann and Kirsch 2003; Kirsch 2015).

In the drama Faye Wong plays the Chinese designer Faye Lin who dreams of studying with a famous Japanese designer, but is unable to obtain a visa.[13] A friend helps her forge her name onto the *koseki* (family register)[14] of Akira Suzukake (Kiichi Nakai) who is about to marry his fiancée Emi (Yukie Nakama). Akira, who does not know that his family register has been manipulated, is surprised to find a wife by the name of Faye Lin recorded in it on his wedding day. As advised by a friend, who happens to work at the family register department, Akira goes looking for this mysterious woman in order to divorce her. In the meantime, his friend promises that he will submit the documents from his marriage to Emi after Akira has obtained a divorce from Faye. Because Akira does not want to hurt Emi, he fakes their marriage on the premise that soon he will be divorced and then will be properly married. Thus, Akira lets Emi believe that they are actually married, while in reality he is married to a woman he has never met.

When Akira finally finds Faye, she does not want a divorce, because she cannot obtain a visa legally otherwise. Though reluctant at first, Akira agrees to stay married, so that she can pursue her dream of becoming a designer. As he is a photographer, and thus a creative person himself, he understands her motives far better than he might at first admit. In line with the common depiction of Asian energy (Iwabuchi 2002; Gatzen and Gössmann 2003, Kirsch 2015), Faye goes on to pursue her dream.

13 Faye is supposed to originate from the People's Republic of China.

14 In Japan the wedding ceremony has no legal standing; the marriage is legally complete when the groom transfers his bride's name onto his family's *koseki*. Thus, the family name is always handed down through the male side and hyphenated names or different names are not possible.

Having decided to support Faye, Akira has to continuously lie so that Emi does not realize that she is not officially married to him. Faye, in the meantime, goes on to realize her ambitions – but the designer with whom she has dreamed of working does not keep his promise to help her obtain a working visa. Thus, she still relies on Akira in order to stay in Japan. Through the course of the series, Emi and Faye meet and become friends while Akira and Faye fall in love. After Emi realizes what is going on, she selflessly relieves Akira of his promise to marry her, so that he is free to be with Faye. Emi, however, is not to be pitied, her heart has wandered away at the same time as Akira's and she has a new lover waiting.

What is important here is not how Faye is depicted (she changes from a selfish, scheming character to an altruist who would sacrifice anything for others),[15] but how she is marked as Other. Although it is stated clearly at the series' beginning that she is Chinese, not much reference is made to her actually being Chinese. She rarely speaks Chinese, but uses English or occasionally Japanese – although only after she seems to have been in Japan for a while or in key scenes for which it is vital that the audience be able to understand the dialogue without reading subtitles. Since Akira does not understand either Chinese or English, and both have to be subtitled for the Japanese audience, the choice of language should not have mattered. Apparently, however, it does make a difference. The frequent usage of English, in combination with her name, Faye – not just the actress' first name, but also a common English name – make her seem more western than Chinese.

As one of the first dramas in the Asia boom, Faye's fluent English seems necessary in order to indicate her difference. In light of the politically volatile relations and the fear of Chinese committing crimes in Japan, this de-construction of ethnicity in Faye's case appears to be a way of marking her as less Chinese – and therefore easier to identify with for a Japanese audience, which, after all, has long been subjected to westernization (Kirsch 2015). This stance is supported by the fact that the Chinese mafia also

15 Although her criminal energy might make her seem a true Chinese, as in the *gaikokujin hanzai* representations, she is not shown as an unlikeable character, but rather as a person stuck in an impossible situation.

briefly appears in the drama. Unlike Faye, the gangsters are stereotypically Chinese, they have long moustaches, wear traditional clothing and have their headquarters in a place filled with *chinoiserie*. Just as Faye seems westernized, they seem as Chinese as possible. Thus, we find both patterns of Orientalism in this drama: a fascination with Faye who, in spite of her westernization, is equipped with an energy seen as lost by the Japanese; and a 'distrust of the bad Orientals' – gangsters, who threaten the main character Akira and who are bestowed with every stereotype imaginable to mark them as Chinese (see also Kirsch 2015).

Additionally, it is significant that all three main characters are artists. Faye is a designer, Akira a photographer and Emi engages in artistic (western style) flower arrangement. They are presented as creative types rather than as two Japanese nationals and one (westernized) Chinese. It is their common artistic dream that makes possible an understanding beyond national boundaries. The three main characters seem to be meeting in a space devoid of specific nationalities and to come to an understanding as artists, despite the fact that Faye's being non-Japanese is important to the plot. This space, moreover, makes a happy ending possible for Faye and Akira (Kirsch 2015).

Considering that the Self always constructs its Other, Faye's character is designed to make her familiar enough to ensure a sufficient degree of audience identification and alien enough to make the plot possible. In de-emphasizing her Chinese-ness and conferring on her something more recognizable through her westernization, she appears to be a projection of everything the Japanese crave: Asian energy combined with an international/western flavour. However, Faye is not the only westernized Chinese character that should be considered. In *Honke no yome*[16] (The main household's bride), another TV series aired in the same year, the western stance becomes even more obvious.

16 The *honke* (main household) is the unit whose task it is to uphold the family name and fortune, which requires the birth of a male heir, often putting women under enormous pressure. The families of the non-inheriting younger sons were called *bunke* (branch households); the *honke-bunke* system has no longer has any legal standing but it remains important in rural areas (see Ochiai 1997, 2000; Moon 1992).

## A 'Global Citizen': The Female Protagonist in *Honke no yome*

The television series *Honke no yome* was aired in 2001 by Nihon TV. It starred the Taiwanese actress Vivian Xu, who is fluent in Japanese, and Shima Iwashita, who has become famous for her interpretations of strong traditional women (Kirsch 2015). Vivian Xu plays Nozomi Yamada, a young woman of Japanese and Chinese (Taiwanese) descent who goes to live with her husband's, Shinji, family, a traditional *honke*. Although her husband is only the second son, his elder brother leaves the family to pursue his own dreams and Shinji is asked to take his brother's place as the family heir. Nozomi is not thought fit to fulfil the role of a traditional housewife since she is a journalist and was not raised to do the housework for an extended family; so, Kin (Shima Iwashita), Shinji's step-grandmother and the family matriarch asks her to divorce Shinji. Nozomi, however, does not want to end her marriage, so she decides to stay and submit herself to the work expected of the *honke no yome*. Soon after arriving, she notices that the family's traditional lifestyle does not allow most members to fulfil their innermost wishes and decides to change the lives of everybody in the household.

Unlike Faye in *Uso koi*, who is marked as westernized mainly through the usage of English, Nozomi is depicted as westernized from the very beginning when she introduces herself off-screen as having grown up in the United States. She speaks Japanese, Chinese and English, thus making her seem slightly out of place within the rural Japanese setting, but this enables her to be the perfect person to change various antiquated traditions. Nozomi's hybrid identity as somehow Chinese/Taiwanese, American and Japanese alongside her multilinguality thus is the impetus that turns her into a modernizer within the family without making her seem too progressive or too alien, at the same time, however, some of her non-Japanese behaviour (disliking *sashimi* and being unable to sit on her legs for a prolonged period of time) play again with stereotypes of 'foreigners' in Japan, while enforcing her difference (Kirsch 2015).

Nozomi's character thus accords with Mathews' (2000) conclusions about the younger generation in Hong Kong, who seem to choose their identities not necessarily from within national boundaries, but who make use of global cultural commodities to build their identities. Mathews notes that:

> Hong Kong in the late 1990s constitutes one of the world's most heterogeneous cultural environments. Younger people, in particular, are fully conversant in transnational idioms, which include language, music, sports, clothing, satellite television, cybercommunications, global travel and ... cuisine. It is no longer possible to distinguish what is local and what is not. In Hong Kong ... the transnational *is* the local.
> (Mathews 2000: 136, italics in the original)

Nozomi is represented as having made similar choices among the cultures on offer to her; she appears difficult to place within national boundaries, but it is exactly this heterogeneity that permits the audience to identify with her, since she simultaneously is not too foreign and yet still somehow Japanese (Kirsch 2015). In short she is a stranger in Simmel's (1968) sense: close and yet distant. It is again her westernization and lack of a too overt Chinese identity which integrates her into the family, as well as endearing her to the audience. Although she might represent a growing minority within Japan, Nozomi is not the only fictional character of bicultural descent on television, in the drama *Kowloon de aimashō* (*See you in Kowloon*), another young woman of Japanese and Chinese descent appears, while again, the West plays a crucial role.

## Westernization as a 'Curse': *Kowloon de aimashō*

The TV series *Kowloon de aimashō* was aired in the summer of 2002 by the private broadcasting station Asahi TV. One of the main characters is played by Tsuyoshi Ihara, who became internationally famous after portraying Colonel Nishi in *Letters from Iwo Jima* (2006). In the series his co-stars include actors such as Yuriko Ishida, Kyōko Hasegawa and the singer

Ryūichi Kawamura. The drama deals with several love triangles, one of which involves a woman of Japanese and Hong Kong Chinese descent. This young woman, Jasmine, has followed the Japanese love of her life to Japan, only to find out that he is cheating on her; however her true love eventually wins him back. In contrast to Jasmine (played by the Japanese actress Kyōko Hasegawa), the Japanese female character Kaoru has a rather troubled love life. She is having an affair with her boss Akazawa, knowing that he is married. When Kaoru falls in love with one of her colleagues, Shutarō Kamisanjō, she is at first unable to end the relationship with Akazawa and has two lovers at the same time.

Thus, Kaoru's irregular love life is contrasted with Jasmine's fervent belief in true love. It is also noteworthy that Kaoru is a career woman who would do anything to be successful – including sleeping with her superior – whereas Jasmine's only ambition is to earn enough money to live close to her boyfriend, marry him eventually and be happy. As the plot evolves, it becomes clear that Kaoru is increasingly unhappy with her decision to pursue a career, and she eventually quits her job. Only then is she able to get rid of Akazawa and to be truly happy with her lover Kamisanjō.

Here, unlike the other dramas, it is not the Chinese character who appears as westernized, but it seems to be a case of Japanese westernization gone wrong for Kaoru and Akazawa. Kaoru is ambitious and modern in every possible meaning of the word: she seems to have lost her sense of values, having consciously entered into a relationship with a married man, which, according to the values outlined in *nihonjinron* literature, should never occur in a society which places so much emphasis on family. Given this, she not only appears as ambitious, but also as selfish and individualistic, going her own way – all attributes which are commonly associated with the West. In addition, to mark her as even more western, she lives in a western-style apartment without any Japanese commodities.

Akazawa, her married lover, seems to be her male counterpart. He, too, thinks mainly of himself, putting the company's interests well below his own; again making him a representative of western attitudes which do not emphasize the well-being of the group. Like Kaoru, he does not think about his family when he starts the affair and like her, he lives in a totally westernized house.

While Kaoru is contrasted to Jasmine, who because of her decent behaviour is portrayed as a role-model for Kaoru; Akazawa is countered by Kamisanjō, who appears as the perfect collectivist, placing emphasis on the interests of the group rather than his own. When Akazawa makes a grave mistake which could possibly cost the company a fortune, Kamisanjō comes up with a plan to save them, but as he presents the idea, he claims that it was actually Akazawa's and not his own. When Akazawa confronts him about this (he seems unable to understand the concepts of selflessness or care about the group's welfare), Kamisanjō explains that it was after all in the company's best interests, which should be valued above his own. Thus the two men play out stereotypical *nihonjinron* roles. Akazawa represents the individualist and selfish West, whereas Kamisanjō represents the group-oriented Japanese. Here, the same westernization that has enabled Nozomi in *Honke no yome* and Faye in *Uso koi* to integrate into Japan, becomes a curse and is constructed as a less desirable way of living for the Japanese characters. The message of this drama would appear to be that westernizing to the extent that being Japanese is forgotten will lead to problems within society. In this manner, Kamisanjō and Jasmine present a way out of these self-inflicted social problems (Kirsch 2015).

However, while Nozomi appears as a more international character with her diverse background and multilinguality, Jasmine, who is played by a Japanese actress, is somehow rendered more Chinese than Nozomi or even Faye. Linguistically, her Japanese is halting and she has an accent, all of which gives the impression that she is not a native speaker, she does not resort to another language (such as English) as Nozomi and Faye do, as this would probably reveal her as being Japanese, since it is easier to imagine a foreign accent in your mother tongue than to pretend having another foreign accent in a foreign language. In addition to her linguistic inability to speak 'flawless Japanese', her belief in *feng shui* further underlines her being Chinese. In contrast, the actress playing Nozumi, Vivian Xu, does not need to be turned into an Other – as she is famous as being a Taiwanese actress and the drama was also promoted as starring her. Thus, it is not entirely a surprise that in comparing the two characters directly, Nozomi seems to be more Japanese than Jasmine, which is also underlined by her Japanese name, seemingly in need of fewer alienating signifiers. Interestingly, both dramas use the Otherness of these two characters somehow differently but also similarly nonetheless – being not-too-Japanese and not-too-Chinese integrates and excludes them at the same time (Kirsch 2015).

## The Curses and Blessings of Internationalization and Westernization

I have argued that the dramas depicting Sino-Japanese encounters at the turn of the millennium were not solely about the two countries in question, but the West also played a crucial role. In the examples that solely presented westernized Chinese characters, it is their westernization that enabled them to communicate in a more open way with the Japanese. This allowed for Chinese-ness to be de-emphasized and therefore, paradoxically, served to integrate them into a Japanese surrounding, be it as a modernizer like Nozomi in *Honke no yome* or as an example for a successful intercultural relationship as in *Uso koi*. The dramas analysed thus seemed to convey the notion that China needs to westernize itself in order to be fully understood by Japan (Kirsch 2015).

However, in *Kowloon de aimashō*, the process of westernization seems to have gone wrong for the Japanese characters, to the extent that their Japanese-ness is de-emphasized. They seem selfish and individualist beyond comprehension and are thus depicted in a manner often stereotypically associated with the West in *nihonjinron* literature. In contrast to the westernized Japanese characters, the Asian character is constructed as a role-model who does not live such an individualistic lifestyle, but is shown as decent, nice and diligent. The fact that this Asian character is played by a Japanese actress only further underlines the appropriation as she is easy to identify with for Japanese audiences. It is important to note that all the westernized Japanese characters are inspired by their role-model counterparts and return to the 'Japanese way' (Kirsch 2015).

All these dramas explore an accord between Japan and Asia and, to a certain extent, relocate Japan: moving it from its rather detached position towards the rest of Asia.[17] Here, westernization, or countering it, emerges as the crucial point. Just as the West is looked at ambiguously in *nihonjinron* writings, it is appropriated in an ambivalent way in these dramas. While westernization helped the Chinese characters to integrate into Japan and to render them less Chinese, it seemed to be a curse for the Japanese characters. They were either too individualist or stuck in a rut and unable to make progress. One of the key topics of Japanese post-war history, namely the alignment of Japan with the West (along with a resulting westernization) was shown as having had a negative impact on the Japanese and in trying to convey this message, *nihonjinron* tropes were used, making these dramas a visual form of *nihonjinron* (Harvey 1995; Matsuo 2003).

In this sense, in those dramas, the Other, like a shape-shifter, was cast in whatever mould the plots required: as a threat to a national identity or as a possible means to a better understanding. Interestingly, it is only the West which is appropriated in this manner. While in the Japanese case, internationalization always meant westernization and a threat of to the Self; Chinese identities became more acceptable if coupled with another Other. Thus, if something is already alien, western qualities were used to render that alien-ness less strange (Kirsch 2015).

The Chinese characters or, more generally, the idea of returning to Asia, were used to show the Japanese a way out of their self-inflicted westernization. Thus, in these dramas at the turn of the millennium, the 'totalizing triad' in the formation of post-war Japanese identity continued to be important. Considering the fact that TV dramas help to narrate the identities of the imagined community that is Japan, the positioning of that identity in the dramas is intriguing. While the West helped Asia to integrate into Japan, the Japanese still needed to free themselves from too strong of a western influence. At that time, the early millennium and during the Asia boom, Asia served as the new role-model that helped overcome excessive

17 In a sense these dramas also convey the cornerstones of the Asian Values Debate (see Kirsch 2015).

westernization. In that respect, the dramas at the dawn of the new millennium are crucial in their way of dealing with transcultural encounters, as they negotiated crucial questions raised by Japanese post-war identity – and showed ways of relocating Japan post-Cold-War. Through their common westernization, the Asian characters are shown as being able to integrate themselves in Japan and through a Japanese rejection of the West, Japan is able to return to Asia. Either way, the 'totalizing triad' that Iwabuchi writes about, is not dissolved and we are left with the question of which is the more important factor in the creation of a Japanese identity within the global context.

## References


Anderson, B. (1994). *Imagined Communities. Reflections on the Origin and Spread of Nationalism*. London: Verso.

Antoni, K. (1996). '*Japans schwerer Weg nach Asien. Geistesgeschichtliche Anmerkungen zu einer aktuellen Debatte*'. In I. Hijiya-Kirschnereit (ed.), *Überwindung der Moderne? Japan am Ende des zwanzigsten Jahrhunderts*, pp. 123–45. Frankfurt a. M.: Suhrkamp.

Bauman, Z. (1997). *Flaneure, Spieler und Touristen. Essays zu postmodernen Lebensformen*. Hamburg: Hamburger Edition HIS Verlagsgesellschaft.

Befu, H. (2001). *Hegemony of Homogeneity*. Melbourne: Trans Pacific Press.

—— (1992). 'Symbols of Nationalism and Nihonjinron'. In R. Goodman and K. Refsing (eds), *Ideology and Practice in Modern Japan*, pp. 26–46. London: Routledge.

Bhabha, H. K. (1990). *Nation and Narration*. London: Routledge.

Creighton, M. (1997). '*Soto* Others and *Uchi* Others. Imaging Racial Diversity, Imagining Homogeneous Japan'. In: M. Weiner (ed.), *Japan's Minorities*, pp. 211–38. London: Routledge.

—— (1995). 'Imaging the Other in Japanese Advertising Campaigns'. In J. G. Carrier (ed.), *Occidentalism. Images of the West*, pp. 135–60. Oxford: Oxford University Press.

Faulstich, W. (1996). 'Zwischen Exotik, Heil und Horror. Das Fremdartige als Dramaturgie von Kultur'. In E. W. B. Hess-Lüttich (ed.), *Fremdverstehen in Sprache, Literatur und Medien* (4), 413–27. Frankfurt a. M.: Peter Lang.

Gatzen, B. (2002). 'NHK's Visions of Asia'. In: *Images of Asia in Japanese Mass Media, Popular Culture and Literature*, papers presented at the 2nd International Convention of Asian Scholars (ICAS 2) in Berlin, Germany, 9–12 August 2001. *Electronic Journal of Contemporary Japanese Studies*: <http://www.japanesestudies.org.uk/ICAS2/Gatzen.pdf>.

—— and H. Gössmann (2003). 'Fernsehen als Spiegel und Motor des Wandels? Zur Konstruktion von China und Korea in japanischen Dokumentarsendungen und Serien'. In H. Gössmann and F. Waldenberger (eds), *Medien in Japan. Gesellschafts- und kulturwissenschaftliche Perspektiven*, pp. 244–80. Hamburg: Institut für Asienkunde.

Gössmann, H. (2008). '*Die Macht der Fiktion. Zum Einflusspotential von Fernsehdramen in Japan*'. In M. Jäckel and M. Manfred (eds), *Medienmacht und Gesellschaft. Zum Wandel öffentlicher Kommunikation*, pp. 83–97. Frankfurt: Campus Verlag.

—— (2000). 'New Roles for Men and Women? Gender in Japanese TV Drama'. In T. J. Craig (ed.), *Japan Pop! Inside the World of Japanese Popular Culture*, pp. 207–21. Armonk: East Gate.

—— and G. Kirsch (2003). '(De)Constructing Identities? Encounters with 'China' in Popular Japanese Television Dramas'. *Papers of the International Conference Media in Transition 3: Television at the Massachusetts Institute of Technology*. <http://web.mit.edu/mit3/papers/goessmann.pdf>.

Gudykunst, W. B. and T. Nishida (1994). *Bridging Japanese/North American Differences*. Thousand Oaks, CA: Sage Publications.

Hahn, A. (1994). 'Die soziale Konstruktion des Fremden'. In W. M. Sprondel (ed.) *Die Objektivität der Ordnung und ihre kommunikative Konstruktion*, pp. 140–63. Frankfurt a. M.: Suhrkamp.

—— (1992). 'Überlegungen zu einer Soziologie des Fremden'. *Simmel Newsletter* 2(1), 54–61.

Harvey, P. A. S. (1995). 'Interpreting Oshin. War, History and Women in Modern Japan'. In B. Moeran and L. Skov (eds), *Women, Media and Consumption in Japan*, pp. 75–110. ConsumAsian Book Series. Honolulu: University of Hawai'i Press.

Hayashi, K. (2005). '*Dorama 'Fuyu no Sonata' no Seijiteki Narumono*'. *Tōkyō Daigaku Daigakuin jōhō gakkan kiyō jōhō kenkyū*, 69 (March): 55–81.

—— and E. J. Lee (2007). 'The Potential of Fandom and the Limits of Soft Power: Media Representations of the Popularity of a Korean Melodrama in Japan'. *Social Science Japan Journal*, 10: 197–216.

Herbert, W. (2002). *Japan nach Sonnenuntergang. Unter Gangstern, Illegalen und Tagelöhnern*. Berlin: Dietrich Reimer Verlag.

Hijiya-Kirschnereit, I. (1988). *Das Ende der Exotik*. Frankfurt a.M.: Suhrkamp.

Hook, G. D. and M. Weiner (eds) (1992). *The Internationalization of Japan*. London: Routledge.
Huntington, S. P. (1993). 'The Clash of Civilizations?' *Foreign Affairs* 72(3): 22–49.
Iwabuchi, K. (2002). *Recentering Globalization. Popular Culture and Japanese Transnationalism*. Durham, NC: Duke University Press.
Japan Tourism Marketing Co. (updated 2014). 'Historical Statistics – Japanese Tourists Travelling Abroad'. <http://www.tourism.jp/en/statistics/outbound/>.
Katō, S. (1992). 'The Internationalization of Japan'. In G. D. Hook and M. Weiner (eds), *The Internationalization of Japan*, pp. 310–16. London: Routledge.
Kirsch, G. (2001). *Das Bild ausländischer Figuren in der japanischen Fernsehwerbung*. Unpublished MA thesis, Department of Japanese Studies, University of Trier.
—— (2007). 'Spiritual Healing in China? Encounters with the People's Republic of China in Japanese Cinema and TV Drama'. In M. Schönbein and S. Köhn (eds), *Facetten der japanischen Populär- und Medienkultur* 2, pp. 45–68. Wiesbaden: Harrassowitz Verlag.
—— (2015). *Contemporary Sino-Japanese Relations on Screen. A History, 1989–2005. SOAS Studies in Modern and Contemporary Japan*. London: Bloomsbury Academic.
Mathews, G. (2000). *Global Culture/Individual Identity. Searching for Home in the Cultural Supermarket*. London: Routledge.
Matsuo, Y. (2002). *Terebi Dorama o Yomu. Eizō no naka no Nihonjinron*. Tokyo: Metropolitan.
Martinez, D. P. (1998). 'Gender, Shifting Boundaries and Global Cultures'. In D. P. Martinez (ed.), *The Worlds of Japanese Popular Culture: Gender, Shifting Boundaries and Global Cultures*, pp. 1–18. Cambridge: Cambridge University Press.
Miller, R. A. (1982). *Japan's Modern Myth. The Language and Beyond*. New York, Tokyo: Weatherhill.
Moon, O. (1992). 'Confucianism and Gender Segregation in Japan and Korea'. In R. Goodman and K. Refsing (eds), *Ideology and Practice in Modern Japan*, pp. 196–209. London: Routledge.
Ochiai, E. (1997). *The Japanese Family System in Transition: A Sociological Analysis of Family Change in Post-war Japan*. Tokyo: The International Library of Japan.
—— (2000). 'Debates over the *Ie* and the Stem Family. Orientalism East and West'. *Japan Review* 12: 105–28.
Okada, Y. (2005). *Terebi Dorama wa Suki datta*. Tokyo: Iwanami Shōten.
Penn, W. (2003). *The Couch Potato's Guide to Japan: Inside the World of Japanese TV*. N.P.: Forest River Press.

Prieler, M. (2006). 'Japanese Advertising's Foreign Obsession'. In P. Lutum (ed.), *Japanizing: The Structure of Culture and Thinking in Japan*, pp. 239–71. Berlin: Lit-Verlag.

Revell, L. (1997). '*Nihonjinron*. Made in USA'. In P. Hammond (ed.), *Cultural Difference, Media Memories. Anglo-American Images of Japan*, pp. 51–81. London: Cassell.

Richter, S. (2004). 'The History Textbook Controversy as an Indicator of National Self-Reflection'. In G. Foljanty-Jost (ed.), *Japan in the 1990s: Crisis as an Impetus for Change*, pp. 213–24. Berlin, Münster: Lit-Verlag.

Rogge, J. U. (1987). 'Kritik an der Wirklichkeit. Zur psycho-sozialen Funktion von Fernsehserien'. In H-J. Schmitz and H. Tompert (eds), *Alltagskultur in Fernsehserien*, pp. 73–94. Hohenheimer Protokolle, Vol. 24. Stuttgart: Akademie der Diözese Rottenburg.

Russell, J. G. (1994). *Nihonjin no Kokujinkan. Mondai wa 'Chibikuro Sanbo' dake de wa nai*. Tokyo: Shinhyōron.

Said, E. (1978). *Orientalism*. New York: Random House.

Schilling, M. (1999). *Contemporary Japanese Film*. New York: Weatherhill.

Simmel, G. (1968). *Soziologie. Untersuchungen über die Formen der Vergesellschaftung*. Berlin: Duncker & Humbolt.

The Statistical Yearbook of Japan. (2008). 'Registered Foreigners by Nationality'. <http://www.stat.go.jp/english/data/nenkan/1431-02.htm> accessed 25 August 2008.

Thornham, S. and T. Purvis (2005). *Television Drama. Theories and Identities*. Houndmills: Macmillan.

Tōkyō News Mook (1994). *Terebi Dorama Zenshi 1954–1994*. Tokyo: Tōkyō News Tsūshinsha.

Washinton, J. H. (ed.) (2006). *Gaijin Hanzai Ura Fairu*. Tokyo: Eichi Mook.

Weiner, M. (ed.) (1997). *Japan's Minorities. The Illusion of Homogeneity*. London: Routledge.

Wieczorek, I. (2001). 'Kontroversen um den Yasukuni-Schrein. Kriegsmahnmal oder Symbol eines japanischen Nationalismus'. *Japan aktuell* (August): 382–89.

Yamamoto, R. (2004). 'Alien Attack? The Construction of Foreign Criminality in Contemporary Japan'. In A. Germer and A. Moerke (eds), *Grenzgänge. (De-) Konstruktion kollektiver Identitäten in Japan*, pp. 27–57. Japanstudien. Jahrbuch des Deutschen Instituts für Japanstudien Vol. 16. München: iudicium.

Za Terebishon (ed.) (2004). *Rendorama 10nenshi. Dorama Akademī Shō 10nen Purēbakku. The Television Drama Academy Awards*. Special Issue of *Za Terebishon*. Tokyo: Kadokawa Shoten.

BRUCE WHITE

# Japanese Reggae and the Def Tech Phenomenon: Global Paths to Intra-Cultural Pluralism

## Introduction

This chapter sets out to demonstrate how youth communities of both reggae and Def Tech artists and fans seek to transform their sense of whom and what their society can potentially include and represent. More specifically, it will focus on how both fringe and mainstream artists and fan communities attempt to diversify the concepts, ideas and images through which Japanese cultural membership is imagined and legitimized. I will argue that the people within this study are agents attempting to transform the perceived symbolic structure of their societies. Given the specificity and focus of their agency I will describe these individuals as active change seekers. Unlike some of their generational counterparts (*hikikomori* [those avoiding social contact] or *otaku* [obsessive hobbyists] to name two such groups), who are often portrayed as victims of social isolation and/or passive agents of change, I see these fans as attempting to change their society in order that it might more broadly represent what they perceive to be their legitimate and morally justifiable values and concerns. For the people I describe, their sense of isolation leads to a need to connect and transform key symbolic bases of their society.

Nevertheless, there are differences in the way fringe and mainstream groups attempt to make Japan a more multi-cultural nation-state and there is even disparity between individuals within the same group. In this chapter, however, I pick up the themes of this book and am concerned with emphasizing how the global importation of ideas and images provides the material from which the attempts to create cultural diversity in Japan are lent authenticity. These global images act as conceptual cores where 'foreign' forms of representation and values are reassembled, enabling them to affect the wider cultural symbolic structure in ways that create new spaces for identity-making and which support claims for legitimacy.

The research on which this chapter is based was gathered over approximately three years of multi-sited fieldwork (2006–11) in Tokyo and Kyoto. The research focused primarily on extended ethnographic interviews with key individual fans, both at the sites of musical performance and in follow-up interviews in a variety of public locations. For key informants like Ken (below), I would often interview informants' immediate circles of friends and family and attempt to observe them in multiple contexts outside the musical fandom circles which are described herein.

## Reggae in Japan

The casual observer might have difficulty discerning Ken's true passion, given his closely shaved head, quiet manner and job as a care worker in Tokyo. Ask this 35-year-old Japanese father of two what drives him across many areas of his life and he will cite a new religious movement whose God is a former Ethiopian emperor and whose key philosophy invokes the power of resistance over oppression. 'Reggae, and reggae's message, is kind of, well, *my whole world*. I probably attend every major event in the Tokyo area.' Ken smiles and shows me a photo of his wife and son on his mobile phone. 'I met my wife at a reggae event and now we take my five-year-old son and baby girl to every event. We are a reggae-mad family!'

In our first interview Ken shared with me his thoughts of how reggae was first introduced to Japan:

> Reggae has been around in Japan since the 70s and some of the older and seriously committed guys in the scene are still around. Years ago they were the first to travel to Jamaica, see Bob Marley in concert there, and then in Japan on his visit in 1979.

They were the ones who first really absorbed the original reggae spirit (*supirittsu*). Some of these guys are now in their 60s, and the most committed of them are still proper 'Rasta Men', vegetarians with dreadlocks who still walk the streets of Tokyo looking like the archetypal Rasta, attending the Jamaican Embassy's annual festivals and turning up at the impromptu gatherings of reggae artists and fans at Tamagawa to bang drums and talk about the importance of the Rastafari movement.

These older Rasta Men have a proximity to the original idea of what reggae was about, and serious fans like me and my friends respect them. Those of us in our 30s for instance still understand the spirit behind the music, even if we don't dreadlock our hair, and still eat meat, consume the products of society, or work in proper jobs as 'members of society' (*shakaijin*).

Of course, when I say, 'normal lives', I must admit that I am the 'straightest' (*ichiban majime*) person in my group of friends. I mean, I have a fulltime job. There are literally no salarymen (white-collar workers) in my groups of friends, or friends of friends that I know. Now, I am not a salaryman either, but having a 'proper' job (*chanto shita shigoto*) as a care worker for the elderly makes me an exception. All my friends are labourers, truck drivers, unskilled carpenters, gardeners, or such like.

I don't know how much the older crowd see us as sell-outs, because we don't strictly follow the Rasta lifestyle, but many of us still believe very much in the Rasta message, and live as much as is practically possible according to it. For instance we put a lot of value on the family, and children are very important in this, they get all our attention. Love and peace are important concepts, as is a sense of being close to nature – at reggae festivals many people might be barefoot, and children might be running around naked – unusual scenes for Japan, as you will appreciate. So, although we make comprises which allow us to blend in and live comfortable lives in modern society, we see ourselves holding very different ideas about important aspects of life from mainstream people (*futtsu na nihonjin*).

It would be kind of a dream for me to visit Jamaica – to take my wife and kids there for a few weeks, even a few months, or a year or more! Live the life of the local people. I have many friends – especially among the artists and producers and performers – who have visited for long periods, some for a year or more. They live there and master the patois and become so immersed in local life that they often get mistaken for natives by locals! I suppose, though, that I am more an absorber of reggae spirit than I am a producer or artist. I like it that way, those guys work hard to inspire us fans, and all we have to do is receive that inspiration and use it to help us live our lives!

## Importing Global Cultural Products

The existence of reggae's ideology and fandom in Japan, as exemplified by Ken, challenges us to go beyond a simplistic understanding of why global cultural products get imported and the role that such imports play in a local cultural identity's negotiation and production. The assumption might be that the importation of any globally derived image or ideology – reggae into Japan, Japanese managerial practices into US company policy – serves to augment a deficiency or gap in the local cultural representation of itself to itself and/or the outside world. After all, if the image or ideology were already present in the local culture why go to the trouble to import a foreign version?

The Japanese are not, as some critics might say for example, black; how do they have the right to claim affiliation with an African, Jamaican or African American community (see Condry 2006; Cox 2007)? Such critics might add that the attempt illustrates that the Japanese are a naïve homogeneous people who because of their own lack of internal (racial, ethnic) diversity have decided to import/steal someone else's traditions without understanding the historical power abuses/struggles associated with the symbolism they so readily adopt.

Condry (2006) uses West's (1990) phrase 'the new cultural politics of differences' to unpack the problems with the assuming that importing images, ideas and ideologies is an activity which somehow fills a gap, is done for fun, or worse, means that the Japanese don't get it. Instead, Condry (ibid.: 29) emphasizes the idea that imports often serve to address imbalances in *internal* cultural diversity and thus essentialize other cultural or racial images. I will elaborate on this point that global cultural products are imported in order to facilitate such *intra-cultural* manoeuvres. By examining how and why this targeting and manipulation take place, I pay particular attention to the efforts of artists and fans to create *binary oppositions* between various cultural symbols and images in order to achieve specific outcomes.

It is important to note that in no way is reggae the only, or even by any means the largest, music genre to attempt to affect Japan's lexicon of available cultural symbols and images. Following Condry's work on music subcultures in Japan, I hope to draw attention to the attempts to contest and induce change in the status quo in two music scenes: reggae and Jawaiian. My focus is on the attempts of artists and fans to pluralize the status quo and to expand the range of possibilities through which to think about personal expression, societal change and cultural diversity.

Ken provides a good example of how the status quo is rendered as the antithesis of the reggae ideal. Using the term salaryman to represent the 'norm', Ken distinguishes himself and other like-minded fans from mainstream society. This binary opposition is highlighted again through the term *shakaijin*. Ken represents his reggae world as both more conventional than that of the first Rasta Men generation and yet still distinct from the Japanese middle class mainstream. In both examples, opposition is used to represent the symbolic tenets of one group vis-à-vis another, and to comment on these groups' socio-cultural proximity or distance.

In his reference to groups within groups and the divisions which exist within and between them, Ken relies upon an imaginative, conceptual map of the relative position of Japanese cultural symbols and images: an *intra-cultural symbolic lexicon*. His commentary on the position of reggae fans vis-à-vis mainstream society and each other assumes that Ken's interlocutor also has access to this lexicon. Focusing on the arrangement and distribution of symbols in a society – on the intra-cultural symbolic lexicon – can help us to understand the motivations and techniques that individuals and groups use to manipulate intra-cultural imagery in order to affect particular cultural representations' production and legitimacy.

A symbolic lexicon does not dictate how 'words' (symbols) ought to be combined in a whole phrase or communication, but rather sets out the structural limits (definitions) of their existence as concepts to be negotiated in social spaces and encounters. Indeed, the way in which these words/symbols are employed seems highly contextual and improvisational (Machin and Carrithers 1996). Gregg (1998) argues that cultural symbols are arranged/distributed across the cultural imagination, allowing individuals to use a flexible and constantly changing symbolic language. The fact that cultural symbols are distributed, open to manipulation and interpretation, reveals how diverse individuals and groups from the same culture can express themselves differently by assembling or accessing particular symbols in distinctive ways. The phrase *intra-cultural symbolic lexicon* describes this set of distributed social-symbolic resources – a common stock of cultural images from which individuals draw in order to construct and negotiate social meanings and relationships.

## The Japanese Reggae Scene

> There is a natural state of being,
> of using the natural resources of the earth
> to provide for our needs.
> (Japanese male reggae fan, 28)

During the summer months Japanese reggae fans congregate at outdoor festivals and events. Kyushu, Okinawa and Tokyo see the highest concentration of reggae events. Although very large, highly organized gatherings remain common and widespread, the core fan base is best encountered at the informal and free outdoor 'sessions' – events which attract big name celebrities and amateurs who perform together in an 'open mic' environment. One such events takes place in the Kanto (Tokyo and surrounds) region, located next to the Tama river – Tamagawa – on Kawasaki City's border, and it was here that I conducted interviews with artists and fans, while also participating in the various stages of the event during the summer of 2005.

The Tamagawa event brings together an almost exclusively middle fan base (see Condry 2006: 102) of people ranging from 27 to 45 years of age. Given the reputation that reggae communities have for drug use, it might seem surprising that there are a great many children who are brought along by their parents. Indeed, as the music begins in the early afternoon, the passing observer might assume that it is a mix between a huge outdoor extended family barbeque and a bohemian festival of the Woodstock/ Glastonbury variety.

From early afternoon, the music is a selection of recorded pieces carefully arranged by DJs, played through a large PA system. Groups of fans find their own space on the large, open grass riverside. Some come well prepared with folding tables and chairs, draft beer making machines, reggae-style rugs and blankets to hang around their temporary territories. Others are merely content to sit on blue plastic sheets. The children run around happily visiting other family groups and playing. They are encouraged, by their parents and others, to be aware of the music and to dance.

As the afternoon turns to evening, conversations at the event cover a great many subjects. Children are often a topic – the degree to which they appreciate, and are able to dance to, reggae music lends their parents some kudos particularly among childless fans. Jobs, too, are a common topic; and here it was noticeable how many social service and care workers make up the fan base, as do other non-corporate occupations such as artists, craftsman, electricians and teachers. Often related to conversations on occupation are discussions of 'the Japanese' as a mindless consumer group who are unable to understand the impact of their consumer choices. Fans would often point out that their perspective or choices were different from 'the masses' (*taihan na ningen*).

One conversation I had with a carpenter illustrates this point:

> I don't get it. Why don't people understand that local timber, having been subject to same climatic conditions as the area in which one is trying to build would be far superior to any cheaply sourced, probably imported, materials that a large developer would provide? Why are people so lazy, not willing to do research, or take responsibility for these choices?! Of course hiring a large company rather than handpicking professionals will result in substandard, almost certainly unsustainably-sourced, materials and higher costs for lower overall quality! Typical of the Japanese consumer, a type blinded by the convenience of their lifestyle and lazy-thinking.

In a similar vein, conversations on the general pace of life in mainstream Japan, as opposed to within the reggae community, were prolific. My main informant, Ken, said to me: 'Most Japanese live with a high speed, rat-race mentality. Consuming all the time, not giving themselves a chance to stop and think. Not paying attention to the really important things in life – relationships, the environment, just being oneself, getting in touch with who they are.'

For informants who were parents, such comments were often coupled with an explanation of why they thought it important to bring their children with them:

> I want my son to be aware that there are alternative ways to experience the world – that there are ways of thinking that are not dependent on a kind of cut-throat mentality of money, power and influence. Reggae provides a sort of base of mutual human understanding which goes against this shit. We can share this ideology here. We can share our values here. And we can provide a space for our children to pick up the importance of what we have learned, what we believe.

This idea that the world of reggae is one defined against a mainstream society heading towards an unsustainable reality was commonly held regardless of occupation, ideology, life course and parenting practices.

Importantly, my informants did not critique society without putting forth alternative values and ideas. Indeed, reggae was seen to provide the social space in which to reflect and act upon the issues, to instil change through a shared ideology and its transmission to the next generation. Some of these alternatives can be highlighted through describing the rest of the event.

As day turns to night, things enter a different phase; at around 7 p.m. the open-mic phase begins. Next to a make-shift tent housing the PA equipment, a small wooden stage allows a succession of well-known and amateur performers to freely improvise sets surrounded by supportive, dancing, reggae fans. A succession of artists performs sets that are characterized by rapping-preacher style monologues, peppered with Jamaican phrases and intonations. Commonly, each artist refers to the degree to which reggae has changed their perception of the world. It is not unusual to shout out to the audience, in tones more reminiscent of a gospel service than a reggae gathering, questions and phrases such as: 'How many of you out there have had your lives changed by reggae?!' or 'Shout out loud if reggae has changed your life!' Responses to such calls are consistently intensely enthusiastic, sustaining the sensation that one is in attendance at a religious – evangelical – service, rather than an outdoor music event.

Alongside this celebration of the transformative values and identities seen as particular to the reggae community and ideology, this free improvisation contains a good many common themes. These, although overlapping and random in their presentation within and across each artist's set, can be organized into three main ideological categories all of which attempt to address particular perceived deficiencies in the Japanese cultural symbolic lexicon.

The first, most prevalent, theme of all in the free-improvisation sets is the notion that individuals must strive for clear, honest and direct self-expression. This expression should reflect one's true feelings and must be shared as often as possible. Phrases such as 'be yourself', 'be true to yourself', 'open up your heart', 'express yourself', 'express your feelings', 'express yourself directly', 'make your feelings known', or 'be true to your feelings', are peppered throughout the artists' monologues, encouraging the participants to reflect on this theme. I constantly heard comments that harshly depicted the coldness of Japanese communication and considered the consequences of not speaking one's mind: both contributed to the sense that this community perceived itself in philosophical and moral opposition to mainstream forms of emotional expression and communication.

I have already noted how informants attempt to instil the next generation with reggae ideology, and this in itself can be seen to be a form of social agency (Bertaux and Thompson 1993). There is, however, a more focused and individually configured theme of agency evident in the free-improvisation performances. This relates to the second ideological category: encouraging individuals to believe that they can influence change in Japanese (and by inference, any) society. This is to say that the reggae world values of emotional and environmental sustainability, peace, self-expression, etc., are ones that are believed to have change-inducing properties. Through the performances, the fans are encouraged to act on the beliefs that tie them to this community. Here again, some of the phraseology within monologues reflects this: 'go out and change the world'; 'act on your principles'; 'make society a better place'; 'pass on peace'; and 'protect this world from damage'. Likewise, there are the more implicit notions of *doing no harm* by not becoming part of the 'system' or by resisting 'Babylon' (a Rastafari term for the dominant and exploitative mainstream society).

The third category, the notion that people are the same wherever they are from, is difficult to characterize and describe as it appears more in what this community lacks rather than what it directly states. However, it can be said to deliberately act to celebrate local and global cultural diversity through the status ascribed to people on the margins, or periphery, of mainstream regional, racial and historical narratives. As such, Okinawans and people from Hokkaido are given equal status at the local level.[1] Moreover, Jamaicans and 'unconventional' looking foreigners, Asian, black or white with dreadlocks, for example, are given respect and attention. In its trumpeting of an ideology derived from a country that exists in relative cultural and geographic isolation from the centres of western or eastern culture, the reggae community thus places itself outside of dominant world narratives (see White 2006b). This is largely because it perceives mainstream interpretations of diversity to be inadequate, unjust and it thus deliberately sets out to reconfigure the local and global world-map (see also Gerow 2002). Less obvious evidence for this worldview can be found in the lack of dialogue on the specifics of *nihonjinron* or of foreigners as a general category; the lack of any cultural or racial essentialisms in conversations that might otherwise include them; and an absence of referential positioning in a world ranked according to nation, culture, economy or race.

## 'That's Not Me!' Finding Identity through Opposition

Ethnographic observations of how individuals and groups represent themselves to themselves and each other reveal how 'self versus anti-self or me versus not-me oppositions' (Gregg 1998: 135) are used to express a wide range

1 Condry (2006: 214) notes that the rap music scene in Japan also serves to draw 'attention to ideas of race within Japanese society' (see also Steele 1995).

of identities. In the reggae community's representation of these categories – self-expression, agency and diversity – we see how artists and fans work to provide a coherent alternative social membership through positioning key structural symbols in opposition to those that commonly dominate in Japan's intra-cultural symbolic lexicon.

Filial piety, adherence to the group over personal freedom and choice, the expectations of a life course lived in concert with national values that tend to restrict rather than empower the populace; all of these are concepts which reggae artists and fans oppose. Indeed, through anti-me positioning, these mainstream symbols are imagined as a state sponsored symbolic package – a Babylon – that must be resisted and opposed. Through that resistance and opposition, senses of community solidarity are generated and the individual is given a wider range of cultural symbols with which to improvise and expand the project of being Japanese. Observers of Japanese society will recognize these concepts as key structural features that are commonly understood to shape the Japanese intra-cultural symbolic lexicon. They are often found as clusters of images that Japanese individuals attempt to find proximity to or distance from, as is observed, for instance, in the ways in which young people negotiate important decisions in the life course with their parents (Nakano 2006). In the reggae community, however, the efforts to manipulate the position and legitimacy of status-quo images through opposition are particularly intense.

For example, Ken told me during an interview: 'We reggae fans are generally the bums, or the outcasts, of Japanese society' (*hazureta hitotachi, furyō-teki na nihonjin*). Here the image evoked by the terms *hazureta hitotachi* (off the rails, fallen by the wayside etc.) and *furyō-teki na nihonjin* (bums, delinquents, good for nothing Japanese) seem to be invoked as part of a deliberate and somewhat extreme attempt to improve access to a range of previously unarticulated or underdeveloped cultural symbols and images. Indeed, the phrases might be seen to act as 'landmarks' (Machin and Carrithers 1993), that provide individuals with cultural legitimacy, creating a moral and narrative space that symbolically houses those who do not fit into mainstream society. These strategies can also be seen employed by other sub-cultural groups, such as the followers of Def Tech.

## Def Tech (2004–2007)

If the world of Japanese reggae embodies the three key ideological principles (expression, agency, diversity) outlined above, in a community perceived by its fans to operate in opposition to conventional symbolic realms, then Def Tech, a band producing records in the mid-2000s, could be seen to be attempting to shift the not-me symbolic language to the 'me', that is: trying to integrate the messages deeply into, rather than being in opposition to, the structured symbolic language of mainstream society. Such efforts aim to involve a wider generational cohort of individuals and strive to formalize or legitimize the attempts to diversify the intra-cultural lexicon.

It is hard to imagine a sight more global in nature than that of Def Tech, a Japanese man and his Hawaiian-Caucasian band-partner who create a musical mix of Jamaican reggae, American-derived rap and Pacific Islands' melody. For three and a half years between the 2004 and 2007 that is exactly the image to which millions of Japanese were exposed in supermarkets, bars, restaurants, clubs, pubs, on radio and television, in newspapers and magazines. Micro and Shen sold 5 million albums and their innovative Jawaiian Reggae genre, inspired the formation of new bands, won national awards and brought countless Japanese into contact with the vibrant reggae-surfer derived ideology of honest self-expression, sustainable environmentally conscious living, and the open inner search for a 'true' sense of self. Though short lived, it is safe to say that Def Tech was a phenomenon which provided both a legitimization and popularization of these values and images through musical and ideological manipulation. Their success lent the values a recognition that they perhaps had not enjoyed previously to the same extent.

The Def Tech phenomenon relied on a series of interconnected local and global networks that predated the members' very births. Def Tech could not have existed without Bob Marley – Micro's boyhood hero. Nor could they have risen to fame without the creative directors who brought Marley's music to Japan in the 1970s; nor without the Japanese fan base that Ken and his generation of fans represent.

The creative pair who drove this reggae-derived ideology and music reworked the messages arising from the fringe Japanese reggae community described above, and went on to inspire and represent the pluralistic local and translocal lives of Def Tech's more mainstream and younger fan base. The context for Japanese youth's involvement in such an ideology is based on three global cultural phenomena that also underpin the Japanese reggae followers' engagement. The first of these is the need to expand the range of identities available to the self in society. The second relates to the processes through which is engendered a sense of confidence in the legitimacy of youth's values and identities, as opposed to those of adults. This second relates to the third movement: the emergence of a highly articulate and intra-culturally formed cosmopolitanism, or cultural, relativity.

Micro, a Tokyoite and one half of the Micro-Shen duo that is the Def Tech phenomenon, 'fell in love with Bob Marley at three years of age and had mastered Michael Jackson's moon-walk by the age of five' (see deftech.jp). This blend of influences speaks to his role as a facilitator and broker of reggae ideology to mainstream society, predominantly through the popular music market. From their debut in 2004, Shen and Micro captured the imagination of hundreds of thousands of young people around the country, mostly by organizing and packaging the reggae sound and ideology in a less overtly revolutionary – and thus more popularist – Hawaiian/Jawaiian framework than that of Rastafarianism.

Shen is Hawaiian and provided what he termed a Pacific Island sound to the reggae phraseology. In the transition, the music and message are largely stripped of any sign of 'unseemly' Rastafarian roots such as drug or sexual references, although much is retained including a Reggae/Jamaican syntax and the principles of sustainability and renewal. Def Tech renders reggae ideology with Surf rather than Spliff,[2] a process which allows a much larger proportion of people to join forces with its message.

2 Note that I found very little actual marijuana usage amongst the middle fan base of the reggae community.

Def Tech also caught another wave, to paraphrase the title of their third album.[3] If the main fans of the Japanese reggae community are in their 30s, Def Tech reached out to people in their teens and twenties who were already integrating these messages into their lives and thus were well aware of the issues that the ideology addresses. Indeed, Def Tech provided a kind of symbolic validation for the younger generation's own social ideals, values, worldviews and identities, manipulating the symbolic lexicon to allow them to find a sense of social agency and participation.

In contrast to the variety of different events on offer for reggae fans around the country, Def Tech fans could only attend the band's rather irregular concerts. The live shows attracted the younger, and more dedicated, section of an already youthful fan base. They were organized by professional event companies in large concert halls, or similar venues, and attracted numbers in the thousands; typical, perhaps of most popular music concerts. The set I observed, at Kyoto Kaikan in May of 2006, was impressive for the enthusiasm of the band and the audience, the latter falling very much under the spell of Micro's preacher-like improvisations between songs. In fact, the only advantage in a fan paying to attend a concert rather than purchasing an album, beyond of course seeing the stars in the flesh and experiencing the live atmosphere, was that Micro's enthusiasm was even more contagious on stage where he added to the messages in the music through short monologues. For example, as prelude to the popular track, *Consolidation Song* (2005) a tune largely about the search for meaning in life and relationships and an acknowledgement of a shared humanity, Micro 'preached' the following – without background music, in a voice that rose towards an impassioned crescendo which the music then met:

> We all feel sad sometimes. We all feel happy sometimes. We all feel close to those we love. We need to be able to take our feelings and communicate them. To tell those who we love that we love them. To tell those that we are angry with that we want to make amends. If we can't do this, then what is the point of it all? Be yourself. And

3 *Catch the Wave* (2006), the name of both the song and album, emphasizes the Pacific Island surfing culture influence.

> be with others. Express yourself, and find in others their expression. Let's do this together! Let's do this together!

Through such interchanges with the audience, the messages and ideology in the music were given a special immediacy and relevance that traded on the group's solidarity, the Def Tech presence, as well as their carefully constructed and adapted message. In performances it became clear that Micro was the director of 'the message', and that his enthusiasm for transmitting it lay in his assumption that he could empower youth with a sense of social agency by using the tools necessary to improve the quality of their emotional and interpersonal experiences and relationships. There were three areas that Def Tech targeted – areas strongly related to the principles derived from reggae ideology – that I examine below.

## Agency within Familial and Intergenerational Relationships

The first area is the realm of youth's relationships with their parents – the family dynamics that characterize many of young Def Tech fans' home existence. For a large proportion of these fans, particularly those still residing at home, there was a frustration that what they saw as important was often not recognized nor valued by their elders. For example, many fans to whom I spoke said they felt that their desire to have warmer more intimate relationships where feelings were talked about and all topics were openly discussed, was not shared by their parents. 20-year-old Akiko told me:

> I very rarely get an opportunity to talk frankly with either of my parents about my feelings, or my emotional state. I could say that neither of them have ever opened up to me, and I think because of that I don't really have the confidence to open up to them. When things are really bad then of course I can talk to them, but it is not the same as sharing my feelings and thoughts with my friends, who will listen and sympathize, and try and help me through with kind words.

Often the way that informants felt was seen not to be given any special space for discussion or evaluation at home, and this resulted in feelings of isolation and loneliness – incidentally a state which is cited by countless newspaper and sociology articles as the cause of a variety of youth problems, from *hikikomori* to general delinquency. Def Tech's music helps to encourage these young people to find self-representation through the music – to recognize that they are not alone in their isolation.

> We are brought into the world equal, subject to one time and one face, subject to one life and one death.
> But some things begin to separate us, our environment, family, and the way we are brought up.
> (in English) DNA pain complain insane in the vain brain...
> ('Consolidation Song' (2005), my translation)

Such direct references to the importance of family and upbringing, as well as the subtext that dysfunctional or uncommunicative family environments cause pain, are common in the music. Again, the picture of there being a whole different range of potential environments and family configurations – all of which have the possibility of causing emotional pain and suffering – has an important therapeutic value for those who feel deeply isolated within their own homes.

It is important to note that this search for self-representation as relief for the isolation caused by family relationships is not confined to music. As Kingston (2004: 272) points out, art as a whole can act as a counterbalance to the isolation of contemporary youth. Def Tech perhaps went one step further in actively encouraging fans to find real interaction and cross-peer solidarity. Indeed, Def Tech seemed to explicitly understand that bonding in friendship can help give strength to youth in this position: 'Hold my hand, and you won't be scared, but I suppose I won't always be here, you'll have to do it without me sometimes' ('My Way' (2005), my translation).

Def Tech's lyrics are littered with references to creating partnerships in order to overcome and tackle problems; to find common strengths and purpose. Therefore, as well as providing the artistic material within which young individuals can find themselves represented on an intergenerational stage, Def Tech also provides for the creation of intra-generational solidarity.

As seen in the above short extract the band was not overly romantic or optimistic. Peer-group solidarity is essential for initial empowerment, but ultimately one must use it as a vehicle for self-empowerment and individual action.

Peer group solidarity is indeed embodied in the way in which many fans go about creating their social networks, as well as the way in which such networks are utilized as arenas in which their intergenerational problems at home can be discussed in a safe and supportive environment. Fans I spoke to did not generally consume Def Tech music as isolated individuals, but rather as members of small intimate peer groups who shared activities, a love of the music, and approaches to overcoming problems. Kumiko, interviewed after the 2006 live concert, said:

> I probably have five close friends with whom I share all of my problems. We depend on each other for advice – often another friend has been through what we are going through – so we all feel very connected and supported. I don't know what I would do without them! Three of us are here tonight. The other two had work so couldn't make it. We listen to Def Tech over and over again in the car on the way to the beach to surf!

## Inter-Cultural and Intra-Cultural Cosmopolitanism

The second way Def Tech could be seen to represent, and, perhaps to a limited extent, impart a sense of agency relates to the rather complex area of ordering intra- and inter-cultural diversity. First, the obvious deep friendship between Micro and Shen seen on stage and in their music videos is one example of how Def Tech expounded a model of healthy intercultural cosmopolitanism. The tall thin Hawaiian and short stocky Tokyoite were an odd pair, but in the way that they spoke and interacted with one another; worked together in the music's production and performance; and so obviously shared a concern with the ideology and message of their work (put across in both English and Japanese in almost every song); the audience and listeners were brought into the folds of this cross-cultural, inter-language and inter-racial partnership. In so doing, the category of the kind of partnerships that succeed in Japanese society is expanded to include such inter-cultural pairings, and, as a result, Japanese society itself seems momentarily more dynamic and diverse. One did not need to identify with the Def Tech message itself in order to form this impression, it was one brought about merely by Def Tech's existence and mainstream successes.[4]

More importantly, for the fans themselves, this 'visible' and 'audible' (Shen accompanied Micro in perfectly fluent Japanese in many of the choruses) inter-cultural component was merely symbolic of a much more meaningful grassroots intra-cultural cosmopolitanism. This intra-cultural variety can be seen in the way Def Tech's message spoke directly to, for example, the generation gap that exists between the young and their parents, particularly with regard to the conception of who and what Japanese society can and should contain.

Monocultural identities often rely on the notion that all citizens are subject to the same kind of family ideology, community structures and cultural and national experiences. Again Def Tech's message was very much at odds with this idea. Understanding that people are subject to range of different local systems and familial relationships, lends itself to an adaptable worldview which places social diversity at its centre. The acknowledgement of local diversity is clearly vital in promoting the 'foreigners to ourselves' phenomena (Morley 2000; see also White 2006a), as well as a 'degree of consciousness that goes beyond any one situation, an awareness that each moment is embedded with a range of cultural possibilities' (Amit-Talai et al. 1995: 231).

One example, among many, of how Def Tech encouraged its fans to consolidate local and translocal conceptions of diversity, is a short act performed in their live concerts. Micro and Shen cleverly would stage an

4 Def Tech won a prestigious NHK music award (2005), placing them firmly in the Japanese mainstream and elevating them to 'Japanese cultural property' standing; a position, which brought them to the attention of generations older than their core fan base.

argument towards the end of a song themed around the idea of cultural/racial essentialisms' pointlessness. Feigning annoyance at Shen, Micro began by commenting that foreigners can't understand things properly, that they are all individualistic, selfish and think they are superior. Shen counterattacked by saying that Micro was a short and bad tempered Japanese, stereotypical in that he didn't say what he felt until he lost his temper. The argument led to the two making derogatory remarks about racial features such as the size of Shen's nose, and at this point, the argument would suddenly stop and the two singers would face each other and begin to sing about how pointless limited such stereotypes and worldviews are when they are seen in the context of real human relationships.

The message of this powerful performance hit home, not least because many of the acted-out essentialisms were familiar to the young audience as views they would have heard from their parents (see Nakano 1995; Kato 1992). Taking direct aim at the cultural nationalism associated with fans' parents, as well as referring to overcoming a Japanese inferiority complex – a reference to a baby-boomer generation still existing in a mentality which ranks nations according to relative economic or historical power – Def Tech was concerned with the emancipation of youth from narratives which seek to limit their imaginative mobility.

> I think the message in the music is about being honest and upfront and open-minded. There are all kinds of different personalities – of people – in the world but we all share in each other – in our relationships with each other. Being true to ourselves and to each other is what the music's message says to me.
>
> (Def Tech fan, 23, male)

The ability to integrate local experiences of social diversity (various family arrangements, relationships, personalities etc.) with the idea that the same patterns of diversity exist outside the realms of one's immediate face-to-face relationships (i.e. inter-culturally, or inter-nationally), has important implications for the way that those exposed to these efforts may see the Japanese intra-cultural lexicon as representative of their lives.

Just as the reggae world emphasizes the periphery in attempting to do away with the notion that there exists a legitimate stable and monocultural centre, Def Tech challenges a set of internal and external ideas about what it is to be human, stripping the definitions of culture down to a notion of common community. As the track *Consolidation Song* states most clearly: 'It doesn't matter where you come from, let's go as one consolidation song.' In short, Def Tech has taken the key symbolic components that the reggae community was rendering as not-me representations, and integrated them into the very heart of mainstream intra-cultural language as 'me' products.

## Operating in Lexiconical Workspace

> Globalization' is a non-linear, dialectic process in which the global and the local do not exist as cultural polarities but as combined and mutually implicating principles. These processes involve not only interconnections across boundaries, but transform the quality of the social and the political *inside* nation-state societies.
>
> (Beck 2002: 17)

This chapter began with an enquiry into the notion of what it meant to import global images for use in local contexts. The question can now be answered to some extent by seeing that intra-cultural diversification and change are, in themselves, one of the goals of highly orchestrated attempts to shift the position and relative power of mainstream symbols in the intra-cultural lexiconical workspace. This notion of lexiconical workspace highlights the fact that a cultural lexicon is not merely an entity passively accessed, but, for some, is one that can be actively 'edited' and manipulated in 'real-time' space.

Cultural lexicons, like languages, can express any idea and probably have been through countless reassemblages where certain types of symbols have come in and out of usage. In this light, global cultural imports merely serve to help accentuate certain ideas that particular cohorts of people feel may have been 'forgotten', or which need emphasizing or developing. In the story of reggae and Jawaiian music, we see a range of interrelated attempts to cause shifts and transitions in the intra-cultural symbolic lexicon. Artists set out to expand the range or availability of cultural symbols and identities to which individuals have access.

Artists may be seen as 'creative directors', occupying a vital role in the mediation of the symbolic system that determines and guides the production of social meaning and representation. Their efforts can serve to expand the range of cultural symbols, ideas and identities from which individuals choose in order to represent themselves in everyday life. Doing so facilitates a process that on the surface appears to us as a process of globalization, glocalization or grobalization: the importation, or free-flow, of foreign images or ideas into our local societies. It may be more accurate to describe this process as being intensely involved in managing *intra-cultural* shifts, as efforts to expand a society's socio-cultural symbolic landscape or lexicon. A focus on the global may, therefore, be less informative and perhaps even auxiliary to these attempts to *pluralize the local.*

As individuals, the artists and producers of creative industry, in their aim to modify the lexicon, may be related to a need to find for themselves a legitimate form of cultural representation in their own society – to pluralize their locales so that they may find appropriate representation for themselves in that lexicon. Together, then, fans and artists of Reggae ideology, and its subsequent popularization, are important indicators of change to the extent to which cultural and national identities are negotiated by the individual in order to find appropriate/legitimate senses of self-representation. In doing so they also reveal the areas of intra-cultural life that they see as needing development and/or changing. The potential use of the lexicon as a force to mobilize entire cohorts of people, may, in turn, represent an important form of social agency or 'soft advocacy' for social change in Japan and beyond.

More broadly, a focus on how the intra-cultural lexicon is modified helps us to take a more informed approach to the questions of how and why global cultural images are imported into cultural systems of representation. Understanding this can help us to be attuned to the kinds of intra-cultural shifts that are being attempted by the importation. Such imports may not be occurring merely to introduce a concept or ideology into the intra-cultural system, but rather to accentuate, subdue, reinvent, or renegotiate the position of clusters of existing images within the existing symbolic lexicon. This might be being attempted for a wide range of reasons, including: the search for social legitimacy; to expand the breadth of possible self-representation in a lexicon seen as overly dependent on particular symbolic forms; or simply to provide alternative competitive symbolic bases for improvising social life.

Plotting the course of images' and ideologies' global flows from their original location outside the intra-cultural lexicon, to their importation and rendering as local lexiconical markers that (in this case) establish zones of resistance, contestation and change, we see their value in their ability to modify the very cultural symbolic infrastructure that is embedded within everyday consciousness. In moving society towards more diversified models of intra-cultural symbolism, the brokers, producers, artists and consumers – the people and networks which facilitate this process – are, in important ways, adding to range of methods through which individuals can represent or assemble their diverse personalities, worldviews and lifestyles as appropriate and legitimate within their own cultural settings.

## References

Amit-Talai, V. and H. Wulff (eds) (1995). *Youth Cultures: A Cross Cultural Perspective*. London: Routledge.

Bertaux, D. and P. Thompson (eds) (1993). *Between Generations: Family Models, Myths and Memories*. Oxford: Oxford University Press.

Condry, I. (2006). *Hip-Hop Japan: Rap and the Paths of Cultural Globalization*. Durham, NC: Duke University Press.

Gerow, A. (2002). 'Recognizing 'Others' in a New Japanese Cinema'. *The Japan Foundation Newsletter*, XXIX(2): 1–7.

Katō, S. (1992). 'The Internationalization of Japan'. In G. D. Hook and M. Weiner (eds), *The Internationalization of Japan*. London: Routledge.

Kingston, J. (2004). *Japan's Quiet Transformation: Social Change and Civil Society in the Twenty-First Century*. London: RoutledgeCurzon.

Kotani, S. (2004). 'Why are Japanese Youth so Passive?' In G. Mathews and B. White (eds) *Japan's Changing Generations: Are Young People Creating a New Society?*, pp. 31–46. London: Routledge.

Machin, D. and M. Carrithers (1996). 'From "Interpretative Communities" to "Communities of Improvisation"'. *Media Culture & Society*, 18(2): 343–52.

Miller, L. (2004). 'Youth Fashion and Changing Beautification Practices'. In G. Mathews and Bruce White (eds), *Japan's Changing Generations*, pp. 83–98. London: Routledge.

Morley, D. (2000). *Home Territories: Media Mobility and Identity*. London: Routledge.

Nakano, H. (1995). 'The Sociology of Ethnocentrism in Japan'. In J. C. Maher and G. Macdonald (eds), *Diversity in Japanese Culture and Language*, pp. 49–72. New York: Kegan Paul.

Nakano, L. and M. Wagatsuma (2004). 'Mothers and their Unmarried Daughters: An Intimate Look at Generational Change'. In G. Mathews and B. White (eds), *Japan's Changing Generations*, pp. 137–54. London: Routledge.

Ohnuki-Tierney, E. (1993). *Rice as Self: Japanese Identities through Time*. Princeton, NJ: Princeton University Press.

Steele, W. M. (1995). 'Nationalism and Cultural Pluralism in Modern Japan: Soetsu Yanagi and the Mingei Movement'. In J. C. Maher and G. Macdonald (eds), *Diversity in Japanese Culture and Language*, pp. 27–48. New York: Kegan Paul.

White, B. (2006a) 'The Local Roots of Global Citizenship: Generational Change in a Kyushu Hamlet'. In G. Mathews and B. White (eds), *Japan's Changing Generations*, pp. 47–64. London: Routledge.

—— (2006b) 'Re-Orient-ing the Occident: How Young Japanese Travellers are using the East-West Dichotomy to Dismantle Regional Nationalisms'. In J. Hendry and D. Wong (eds), *Dismantling the East West Dichotomy*, pp. 125–42. London: Routledge.

# PART III

# Technological Connections: Past, Present and Future

HEUNG-WAH WONG AND HOI-YAN YAU

# The More I Shop at Yaohan, the More I Become a *Heung Gong Yahn* (Hongkongese): Japan and the Formation of a Hong Kong Identity

## Introduction

The identity formation of Hong Kong people and its transformation have been much studied; research on the topic has been conducted through the disciplinary lenses of anthropology, psychology, cultural studies, feminism, linguistics and postmodern literary theories, to cite just a few.[1] Notwithstanding differences in approach and/or emphasis, these studies into Hong Kong identity share two major characteristics. First, they tend to utilize 'self-reporting' or 'self-labelling' surveys as their core research method. Second, many of these studies implicitly or explicitly assume that there is a relationship between identity and the properties and characteristics of its signifiers. Ma and Fung (2001), for instance, argue that Hong Kong dwellers' identification with Chineseness was supported by an increasing emotive attachment towards Chinese national icons including the national flag, the national anthem, and the People's Liberation Army (henceforth PLA). Conversely, the increasing resistance among Hong Kong people towards Chinese national icons after the handover of 1997 indicated that they had returned to a local Hong Kong identity (Fung 2001: 598–9). Implied in these arguments is that there is a necessary relationship between Chinese identity and the Chineseness of its signifiers.

1 See, for example the work of Chan 2010; Chun 1996a, 1996b; Fung 2001, 2004; Ma and Fung 1999, 2007; Kim and Ng 2008; Chen, 2008; or Wong, 2004.

With a view toward criticizing this self-labelling survey approach and the assumed necessary relationship between identity and its signifier, we shall comment on the premises of such a method, which presumes a theory of identity first, by initially examining the theory and working back from it to the methodological strategy. This exercise will critique not only the self-labelling survey strategy, but also some of its more bizarre results, such as the constant confusion between identity and opinion. Following Barth (1969), we argue that claiming an identity includes not only self-ascription but, more crucially, ascription by others. Implied in this argument is that, in addition to self-ascription, one has to perform the roles associated with that identity in order to express and validate the self-ascription. Moreover, the maintenance of identity requires continual performance and validation. That is to say, self-labelling without continual performance and validation are, at best, opinions only.

More importantly, as Barth reminds us, identity markers do not follow a descriptive list of cultural differences or features and that the markers that signal a boundary may change over the course of time. For these reasons, Barth argues that the central focus of an investigation of identity lies in the 'ethnic boundary that defines the group, but not the cultural stuff that it encloses' (1969: 14). In other words, the identity marker of 'Hongkongese' does not necessarily have to be something that is from Hong Kong. As we shall show shortly, the identity marker of Hongkongese can be a *Japanese* supermarket, so long as it helps to define the Hong Kong people's ethnic boundaries. Perhaps, then, it is not surprising that studies which draw a direct relationship between a Hong Kong identity and resistance toward Chinese national icons tell us very little about how the identity of Hong Kong people is constructed, or we might even say, assembled.

Even when studies of identity formation and transformation employ other research methods, they often fail to spell out the socio-economic and political changes against which the identity formation or its transformation(s) can be understood. In fact, this specific failure points to a general ignorance of the historical contexts within which a Hong Kong identity is usually made and remade. As we shall show shortly, the current Hong Kong identity has its genesis in the specific historical contexts of post-war Hong Kong society and was interestingly represented through the patterns of consumption that grew up around the Hong Kong subsidiary of a global Japanese supermarket, Yaohan.

The methodological implication is that we can study the emerging identity of Hong Kong people through examining the reasons for the popularity of Yaohan in Hong Kong. We argue that Yaohan was able to garner tremendous interest amongst people in Hong Kong because the cultural significance attached to it concurred with the cultural logic of an emerging middle class identity. That cultural logic largely developed in the rapidly changing environments of the 1970s and is based on an identification with something that is neither eastern/traditional/conservative, nor western/modern/liberal. The 'in-between' image of Yaohan spoke to the same logic, which helps to explain its immense popularity in Hong Kong.

## What is the Identity of Hong Kong People?

The cultural identity of Hong Kong people became an especially popular topic on and before 1997 when Hong Kong was to be handed back to China by the United Kingdom. Ma and Fung, for example, administered a telephone survey in 1996 to explore 'the changing configuration of the Hong Kong identity' in the wake of the sovereign transfer. They reported that Hong Kong people generally displayed a 'dual self-claimed identity of Hongkonger and Chinese' (Ma and Fung 1999: 504), although with a stronger preference placed on the former. More importantly, the distinctive identity of Hong Kong people was supported by a unique ethos (invoked by terms such as westernized, modernized, and so on), and an attitudinal and emotive resistance towards national icons such as the national flag, the national anthem, the PLA, etc. (ibid.).

The academic research on Hong Kong identity continued to proliferate after 1997, although it was concerned more with an identity transformation that was also seen as a form of resistance. Claiming 'self-reporting of one's voice and emotive associations' as a significant socio-political approach that engaged with the question of 'local resistance', Fung (2001: 592) contemplates the change of local identity through surveys conducted between 1996 and 1998. These surveys showed that the percentage of local people using the 'Hong Kong people' label to describe themselves dropped from 25.2 per cent in 1996 to 23.2 per cent in 1997, but rebounded to 28.8 per cent in 1998, whereas those using the Chinese label to describe themselves rose from 25.7 per cent in 1996 to 32.1 per cent in 1997, but plummeted to 24.5 per cent in 1998 (ibid.: 597). These changes in self-labelling were shown to correlate with the local attachment towards national icons, which peaked during the transition as a result of media bombardment, but diminished as soon as the transition was over (ibid.: 598–9). All this thus made Fung conclude that while national discourses during the political transfer penetrated local culture and dampened local identities, as soon as the transition was over, Hongkongers re-adhered to their own label of Hong Kong people in their struggle for cultural autonomy. In other words, the local once again resorted to their Hong Kong identity as a means to resist the national (ibid.: 597).

We respect Fung's (ibid.: 592) moral obligation to 're-construct how the locals imagine their own resistance', but find it difficult to agree with his contention that 'self-labelling is a viable strategy for having an identity.' For we are not sure to what extent changes in self-labelling can be seen as evidence of the changes in Hong Kong identities *per se*. As noted above, Barth (1969: 11) argues that ethnic identity 'identifies itself, and is identified by others.' Here we can see that mere self-ascription (or self-labelling) is not a sufficient condition for the establishment, not to mention the maintenance, of an identity. Recognition by others is also required and, in order to be recognized, one needs to observe two orders of cultural standards:

> (i) overt signals or signs – the diacritical features that people look for and exhibit to show identity, often such features as dress, language, house-form, or general style of life, and (ii) basic value orientations: the standards of morality and excellence by which performance is judged since belonging to an ethnic category implies being a certain kind of person, having that basic identity, it also implies a claim to be judged, and to judge oneself, by those standards that are relevant to that identity (ibid. 1969: 14).

Putting Barth's ideas into our current context suggests that, in order to have a HongKongese or a Chinese identity, one needs to display a specific set of features and follow the basic value orientations pertaining to such an identity. One is also required to continuously express and validate that identity by performing the role prescribed by it (ibid.: 28). For one can only effectively claim a specific identity insofar as, on the one hand, one continuously performs the roles required to signal the identity and, on the other, one's performance is continuously recognized by others. In other words, one cannot acquire an identity by self-ascription only, as Fung assumes.

Fung's unreflective use of the so-called national icons as identity markers likewise appears problematic. As Barth reminds us, cultural standards (as signifiers) do not follow a descriptive list of cultural features or differences. One thus cannot predict in advance which features or differences will be stressed and used as identity markers. Moreover, the markers that signal a boundary may change over the course of time – as may the cultural characteristics of its members. For this reason, Barth (ibid.: 14–15) argues that 'socially relevant factors alone become diagnostic for membership, not the overt, 'objective' differences which are generated by other factors.' This gave rise to Barth's famous contention that the central focus of investigation of identity lies in the 'ethnic boundary that defines the group, but not the cultural stuff that it encloses' (ibid.)

Here, one can see that Fung examines the 'cultural stuff' and most importantly sees it as a diagnostic for Chinese identity. It follows that whether or not national icons such as the national flag, national anthem or the PLA are identity markers of Chineseness is an empirical question rather than a theoretical assumption. Without proving that these national icons are the identity markers of Chineseness, it is very difficult – if not impossible – to argue that Hong Kong people's changing attitudes towards these icons represent a swap of identities from being Chinese to being Hong Kongers, and *vice versa*.

This, in turn, suggests that Fung's major claim that local people once again resorted to the Hong Kong people label to resist the national in the post-1997 era cannot be substantiated. In fact, Chun (2010) argues that on the contrary local people in Hong Kong in the post-1997 era were too ready to side with the new government out of self-interest. Just as media would mute their criticism of mainland China, Hong Kong corporatists would steer clear away from conflict with Beijing and knowingly subdue democratic ideals in order to protect their own vested interests (ibid.: 183–5). This made Chun (ibid.: 185) conclude that 'in the final analysis, who cared about identity, as long as everyone could make ends meet and got what he/she sought, despite the various facades?'

If these national icons have not yet been proved as identity markers, Fung's sketch of the changing attitudes towards these icons is at best an opinion poll of people's attitudes toward those icons. The same is also true of the 'changing contour of the label "Hong Kong people"' (Fung 2001: 600). If the changes in self-labelling are not followed by corresponding changes in performance and subsequent recognition by others, they are nothing more than people's opinions. Barth would probably not disagree with Fung's sketch of the changing attitudes towards national icons, as well as the changing self-labelling, insofar as Fung is not talking about changes in identity as such.

As we can see, the fundamental problem with this kind of opinion poll-like survey is that it does not study identity at all. Its results have nothing to tell us about identity other than its respondents' emotion. Interestingly, Fung (ibid.: 592) himself admits that 'self-reporting' often expresses emotional associations, in which, as expressed in the survey, the media always intervenes.

More importantly, opinions expressed in the survey may also come under the banner of what Goffman (1959) calls the presentation of self. Some respondents might stick to the 'official line' (Becker 1956: 199), while others might 'answer through a filter of what will make them look good' to the researchers (Babbie 1983: 135). Still others might well have as their goal the gaining of acceptance of other participants. This means that Hong Kong respondents could differently represent themselves in accordance with the different contexts within which they are situated. These contexts can range, inter alia, from the era's political climate, to respondents' perception of the researchers' political orientation, to world events such as natural disasters.

We are not suggesting that identities are static and unchangeable. However, a changeable identity is different from a manipulative presentation of self (even if that manipulation may be sub- or unconscious). We are not denying that identities are situational either. However we are not going to give away the integrity of identities. Following Barth, we contend that identity is not something that can be changed easily; nor, as most of the above-mentioned scholars have argued, can identity be so readily switched between Hong Kong and China: identity is not as malleable as post-structuralists would have it. As Sahlins (2000b: 488) remarks with regard to cultural integrity: 'All the same, not everything in the contest is contested ... one cannot legitimately insert a Japanese "voice" in a Sioux Indian ethnography. In order for categories to be contested at all, there must be a common system of intelligibility, extending to the grounds, means, modes, and issues of disagreement.'

Plainly enough, we cannot easily insert a mainland Chinese, not to mention a Japanese, 'voice' into our Hong Kong case. It is therefore of crucial importance not to mix up people's presentations of self as evidence of their 'swapping' identities, but we need to be aware of identification as a process that involves positionality and enunciation (see Hall 1994).

More importantly, if identities are situational, we need to know how they change according to which situations and why. In short, identities are symbolically constituted within a meaningful scheme, which represents one possibility amongst many. That is to say, the identification process has its own logic. We need to know what that logic consists of in order to know how identities change and according to what contexts.

Chun (2009) puts forward an idea that places identity formation firmly in terms of the historical context of a given society, or to use Chun's term, its geopolitics. In other words, identity is nothing more than the result of the unique sociopolitical conditions and a particular society's conceptual influences. Taiwan, Hong Kong and Singapore, for example, have experienced completely different sociopolitical circumstances in the post-war era and, Chun (1996a) argues, they have therefore developed radically different notions of public cultures and identities. Chun (1996b) further explores Chinese identity in Hong Kong, Taiwan, Mainland China and overseas Chinese communities. Suffice it to say that the socio-political power structures in these 'Chinese' societies were different. Chun therefore concludes that the resulting forms of Chineseness manifested in each of these Chinese societies were different.

We share Chun's emphasis on the significance of geopolitics in identity formation. However it is more important to point out, at this juncture, that we disagree with his dismissal of the material impact of the concept of identity on social behaviour. In examining post-war Hong Kong which was marked by changing sovereign powers, Chun (2010: 169) concludes that 'the history of Hong Kong identity(ies) can be seen in some instances more fittingly as a history of hype'. All this may sound much the same as saying that identity is merely a form of false, temporal consciousness, which should have no material impact on individual behaviour. On the contrary we contend here that identities provide a powerful motivation for collective social behaviours. For as an identity is assembled within our social life and experienced as an independent principle of social classification, it can 'motivate social practice and rationalizes the pursuit of individual and collective utilities' (Comaroff and Comaroff 1992: 60). Before we move on to explore how it can motivate social practice and everyday relations, however, let us first take a look at the geopolitics of Hong Kong society against which local identity has taken shape.

## The Geopolitics of Post-war Hong Kong

The founding of the People Republic of China (PRC) by the Chinese Communist Party (CCP) in 1949 put Hong Kong, which had close economic links to China, in a very difficult position. From the start of the Korean War, and the subsequent United Nations embargo on trade with China in the early 1950s, Hong Kong's entrepôt trade, at that time the major economic activity of the Territory, suffered tremendously because China had been Hong Kong's largest trading partner. In addition, the newly established Communist China exerted rigid control over its foreign trade, which made Hong Kong's entrepôt trade decline further (Chiu, Ho and Lui 1997: 30–1). To survive the economic setbacks created by Cold War tensions, as well as to steer Hong Kong away from ongoing national conflict, the Colonial Government chose to transform Hong Kong into a free market port by turning to export-oriented industrialisation (Chun 1996a: 58). This strategy proved to be very successful in the following decade. Manufacturing's share of the GDP increased from 24.7 per cent in 1961 to 28.2 per cent in 1971 and the sector's percentage of total employment went from 43 per cent to 47 per cent (Chiu, Ho and Lui 1997: 52). The rapid growth of the manufacturing sector increased the income of most Hong Kong workers.

Hong Kong manufacturers, however, faced keen competition from other newly industrializing countries, and at the same time suffered from protectionism in their export markets during the second half of the 1970s. Fortunately, China's introduction of open-door policies in 1978 helped to revive the territory's entrepôt trade, while the influx of migrants from China helped to ease Hong Kong's labour shortage. Hong Kong manufacturers were thus able to maintain labour-intensive, low-wage, low profit-margin production in the early 1980s (ibid.: 53–5).

Since the mid-1980s, production costs in the manufacturing sector have risen rapidly because of the soaring property market, labour shortages, and rising wages. The Hong Kong government's ending of the 'touch base' policy[2] in 1980 dried up the constant supply of fresh immigrants whom manufacturers had employed as low-wage labour (ibid.: 55–6). In response, manufacturers relocated their production bases to the mainland in order to exploit the abundant supply of low-wage labour there.

According to Chiu, Ho and Lui this relocation strategy had a twofold consequence for Hong Kong's employment structure. In the general economic structure, there had been a sectoral shift from manufacturing to finance, trading, and other services from the 1970s onward. Meanwhile, manufacturing itself had moved from a production to a commercial orientation. The resulting occupational shift from production to commerce

2 This policy stated that any illegal immigrant from Mainland China who could successfully reach the urban area of Hong Kong would be granted right of abode by the Hong Kong government.

gave birth to a new middle class and changed the class structure of Hong Kong society (ibid.: 71–7).

## The Emergence of the New Middle Class as an In-Between Generation

Lui and Wong (1992) conducted a mobility study in Hong Kong at the beginning of the 1990s. Based on survey data, they discovered that more than 60 per cent of the members of the rapidly expanding service classes did not originate from a service class background. In other words, a good number of people in Hong Kong managed to achieve class mobility through education and hard work during the previous twenty years. Lui and Wong argued that the figure itself proved that a new middle class had emerged in Hong Kong. Given this, we can see that most members of the new middle class were born in the late 1950s and early 1960s, and had grown up in the 1970s. However, the new middle class could only take shape insofar as it began to see itself not as a member of either the PRC or the Republic of China (ROC), but as 'people of Hong Kong'.

Upon the establishment of the PRC in 1949, a large number of capitalists alongside ordinary farmers fled to Hong Kong. Of those Chinese who migrated to Hong Kong during the late 1940s and 1950s, many were supporters of the right wing Kuomintang (KMT) (Tse and Siu 2010: 77). The pro-KMT China faction soon clashed with the pro-Communist China faction in Hong Kong, a thing that the Colonial Government had been attempting to avoid. The intense conflicts between the two factions erupted in the 1956 Riots, and most (in)famously in the 1967 Riots when pro-communist leftists in Hong Kong, inspired by the Cultural Revolution in the PRC, turned a labour dispute into large scale demonstrations against British colonial rule (Chun 1996a: 58; Leung 1996: 145).

In the wake of the 1967 Riots, the Colonial Government began to implement a series of social policies aimed to improve the poor living standard of Hong Kong people, thereby relieving their anger and steering the local people away from further riots. In addition, Governor Murray MacLehose, who arrived in 1972, worked to make society fairer by, for example, establishing the Independent Commission against Corruption (ICAC) to curb the corruption that was a general phenomenon in Hong Kong in the 1950s and 1960s. The ICAC proved to be very successful, not only in curbing corruption but also, through its many educational programmes, in inscribing British/western/modern values of fairness and justice, especially within the new middle class. The new middle class thus absorbed a British/western/modern ethos of fairness and social justice, which had not existed in their parents' generation. However, they could not completely shed the influence of their parents' Chinese/traditional value system, which is why they are regarded as an in-between generation.

## From a Middle-Class Identity to a *Heung Gong Yahn* (Hong Kong People) Identity

We would suggest that the 'in-betweenness' of the new middle class essentially amounts to what Sahlins (1999: 413) calls 'modes of order that are themselves largely imperceptible yet make all the difference between peoples who are perceptibly similar.' This mode of order is a 'structure' which is '[b]uilt into perception, endemic in the grammar, working in the habitus' and works as 'the organization of conscious experience that is not itself consciously experienced' (ibid.). Hong Kong society, its people, and various social phenomena are always described in terms of the blend between East and West, between traditional and modern, and between old and new. The idea that 'Hong Kong is a place where "East meets West", but where "Chinese tradition" still holds sway', as Evans and Tam (1997: 5) have pointed out, has been the way in which people in Hong Kong commonly construct their identity. Evans and Tam also have observed that 'Hong Kong Chinese, when they encounter mainlanders, are able to explain their differences from them by their "westernness", when they encounter expatriates they can explain their differences from them by their "Chineseness"' (ibid.). The same discursive strategy is also adopted by the Hong Kong government to market Hong Kong. Okano and Wong (2004) show that the Hong Kong Tourist Association, which was then under the direct control of the Colonial government, likewise (re)constructed images of Hong Kong by using sets of binary oppositions such as East/West, traditional/modern, rural/urban and so on to promote the territory. More recently, Hong Kong academics have come to understand 'in-between-ness' as constituting the nature of Hong Kong society. As a famous local historian argues: 'studying Hong Kong's role in the Chinese Diaspora as an *in-between place* will enhance our understanding of both the history of Chinese migration and the fundamental nature of Hong Kong society' (Sinn 2009: 245; italics ours).

The in-between-ness of Hong Kong functions in the same way as the Fijian 'living in the way of the land'. As Sahlins points out:

> For instance, the Fijian 'living in the way of the land' (*bula vakavanua*), determined by opposition to the whiteman's 'living in the way of money' (*bula vakailavo*), includes by extension a panoply of practices: chiefship, kinship, a certain generalized reciprocity (*kerekere*), even Wesleyan Christianity, as only Fijians could know it. But then, any one of these, functioning occasionally as an ethnic diacritic (Barth's second type), indexes the class as a total mode of existence, that is by synecdoche or as prototype. (Sahlins 1999: 414)

Hong Kong's in-between-ness, to paraphrase Sahlins, is determined by opposition to mainlanders' Chineseness and foreigners' westernness, while it includes by extension the marketing strategy to promote Hong Kong as a tourist point which is neither East nor West, traditional nor modern, rural nor urban. In-between-ness, we conclude, is a generic signifier for the *heung gong yahn* (Hong Kong people) identity and for the basic value orientations of the new middle class.

The new middle class, unlike its parents, did not identify with the pure Chinese immigrants who tended to regard Hong Kong as a place of temporary shelter. On the other hand, this class gradually became suspicious of the unquestioned welcoming of western culture and values while identifying with Hong Kong, its home. In turn, the in-betweenness identity of the *heung gong yahn* dictated Hong Kong people's consumption culture in the 1980s. A notable example of this consumer culture involves the Hong Kong subsidary of a Japanese supermarket, Yaohan and this requires some explanation.[3]

## 'The More I Shop in Yaohan, the More I Become a *Heung Gong Yahn*'

Yaohan began as a village grocery in Shizuoka Prefecture, Japan. It was December 1930 when Ryōhei Wada, financed by his father-in-law Hanjirō Tajima – Yaohan's founder – opened a branch store in Atami, a hot springs resort town 50 miles west of Tokyo (Wada 1988). The store expanded to include product lines other than groceries as Wada's sons joined the company from the 1950s, thereby developing Yaohan into a general merchandise store or GMS (Wada 1994). In the early 1960s, Kazuo Wada, Ryōhei's eldest son and the senior managing director of the store, went to the United States to study retail businesses. On his return, he persuaded his father to convert Yaohan into a modern retail company. Ryōhei was convinced; he appointed Kazuo as company president and he became chairman (Tsuchiya 1991). Wada then started to build a supermarket chain, opening ten stores within Shizuoka Prefecture from 1962 to 1970.

This brief corporate history shows that Yaohan is not a department store but a supermarket. Japanese department stores and supermarkets differ in three major ways. The first is in the organization of their operations. Supermarkets are self-service operations with chain-style operations, separating merchandising and store operations. Department stores, unlike supermarkets, do not differentiate between these functions (Sato 1978).

3 This case study has been used in different articles to argue rather different points (Wong 1998; Wong and Yau 2015); here we use it to focus on identity making.

The second characteristic of supermarkets is their large number of outlets in residential areas. Department stores and supermarkets are also different in social prestige, a status rooted in their histories and in the physical location of their stores (Larke 1994). Department stores, especially those like Mitsukoshi or Daimaru from the so-called 'kimono tradition', boast longer histories than supermarkets and, within Japanese business generally, a long corporate history tends to be related positively in consumers' minds to quality and prestige. That is to say, a department store generally enjoys higher status than a supermarket.

Additionally, the category of supermarkets is further divided into national supermarkets and regional supermarkets. A national supermarket must, by definition, operate outlets across more than four prefectures and must have a network of outlets in two or more of the following cities: Tokyo, Osaka, and Nagoya (*Nikkei Ryūtsū Shinbun* 1993). The Daiei, Seiyu, Itō-Yōkadō, Jusco (Aeon), and UNY groups are several well-known examples of national supermarkets. Regional supermarkets, on the other hand, are smaller and less well-known. Yaohan only operates in Shizuoka, Kanagawa, Aichi, and Yamanashi Prefectures, and does not run stores in Tokyo, Osaka or Nagoya, classifying it as a regional supermarket. More importantly, just as supermarkets generally are less prestigious than department stores, as a regional supermarket Yaohan has the lowest status within the supermarket category in Japan.

Yaohan's corporate history also reveals that its management and ownership was dominated by the Wada family, which further empowered the company's chairperson Kazuo Wada, who stands out as Yaohan's main historical agent, while his own life represented that of the company as a whole. Kazuo Wada thus had a disproportionate historical effect on the destiny of Yaohan, a fact of which he seemed aware, writing: 'If I change, the world will change too' (Wada 1992: 23).

## Globalizing Yaohan

At the beginning of the 1970s, Kazuo Wada adopted a strategy for survival different from that of other supermarkets in the same region. During the 1960s, some national supermarket chains such as Daiei and Seiyu began establishing stores throughout Japan. This expansion threatened the survival of many regional supermarkets including Yaohan. Instead of merging with other regional supermarkets, or allowing a company take-over, Kazuo Wada chose to go overseas. Yaohan, Wada repeatedly stressed, should keep its own identity. Most of the employees, including the Board Members, opposed the plan at that time, arguing that the company should not spread its already limited capital base by moving into unknown overseas markets. In the early 1970s, Kazuo Wada overcame the opposition from his employees and Board, making his first overseas investment in Brazil (Itagaki 1990). Despite his subsequent withdrawal from Brazil in the second half of the 1970s, Wada continued his overseas strategy, so that by 1995, Yaohan was operating 57 stores in twelve countries and regions.

## Yaohan's Arrival in Hong Kong

In 1984, Yaohan opened its first store in New Town Plaza in Shatin, a new town in the New Territories of Hong Kong. By the late 1980s, Yaohan had established itself as one of the most popular retailers in Hong Kong. Elsewhere (Wong 1998), we have explored the various factors which made the establishment and immense success of Yaohan possible in Hong Kong. Here we just want to demonstrate that Yaohan found favour with people in Hong Kong, because its image matched well the cultural identity of the *heung gong yahn* outlined above.

The image of Yaohan in Hong Kong was established through contrasting itself with other major retailers in Hong Kong as a significant Other. In the 1980s, the local retailing business included Japanese department stores, Chinese department stores (*gwok fo kung sze*), local department stories and local supermarkets. Yaohan differed from other Japanese department stores in Hong Kong in its merchandising policy, locational strategy, and clientele: adopting the business model of Japanese supermarket, locating its chain stores in the shopping centres of densely populated new towns, and supplying daily necessities to local shoppers of the middle and lower middle classes. In addition, Yaohan operated more stores in Hong Kong than did the other Japanese department stores and differentiated itself from the latter in terms of image by means of the business model of Japanese supermarket.

At the same time, Yaohan was different from the *gwok fo kung sze*, local department stores, and supermarkets in terms of its business model and image. In the first place, it offered one-stop shopping to its customers, a service that neither *gwok fo kung sze* nor local department stores and supermarkets could provide. Local supermarkets only sold fresh food stuffs and daily necessities, while *gwok fo kung sze* and local department stores could only provide non-food merchandise. In addition, Yaohan's supermarkets were generally larger and cleaner, offered a wider range of merchandise, and were able to manage merchandise better than did the local supermarkets. *Gwok fo kung sze* did not have their supermarkets and food arcades, and paid less attention to customer service than Yaohan.

The differences in business models between Yaohan and *gwok fo kung sze*, local department stores, and supermarkets led to differences in corporate image. *Gwok fo kung sze* had a strong image of Chineseness, while Yaohan did not; *gwok fo kung sze* were closely related to tradition, while Yaohan was associated with modernity; and local department stores had an image of being conservative and old, while Yaohan represented something new and advanced. Finally, Yaohan was definitely not a western retailer. Yaohan therefore represented something between Chinese and foreign/western, tradition and modern, old and new.

The corporate image of Yaohan, as mentioned above, lay somewhere between West and East, traditional and modern, and old and new. This constituted a happy parallel to the cultural logic of the identity of *heung gong yahn* – a parallel which further made Yaohan into a totem marker for *heung gong yahn* in the 1980s. The company had become a useful identity marker of *heung gong yahn* who were differentiated from Chinese and Westerners in the same way as Yaohan was from other retailers in Hong Kong. Once shopping in Yaohan had become associated symbolically with *heung gong yahn*, more and more Hong Kong people were attracted to shop there as part of signalling and validating their Hong Kong identity. The more people shopped in Yaohan, the more they became *heung gong yahn*.

Thus shopping in Yaohan and the New Town Plaza on Sundays or public holidays became a regular activity for many middle-class Hong Kong families in Shatin. On a normal Sunday morning, the whole family might have dim sum in the Plaza's Chinese restaurants. Afterwards, some members would shop in Yaohan or at other stores in the mall, while others would go to see movies in the nearby mini cinemas, or take their children to the mall's family game centres or funfair-style shops. If they got hungry, they could eat in Yaohan's food arcade or other fast food restaurants, and at dinner time, some would purchase food in Yaohan's supermarket and return home to cook dinner, while others would have their dinner in a Chinese restaurant. This was a typical Sunday outing for many middle-class families in Shatin in the 1980s, a new culture of consumption that continues to this day.

This new consumer culture also moulded the lifestyle of the second generation of the new middle class, who were socialized into their parents' culture of consumption. One of its major characteristics was that 'going shopping' became a major activity. This second generation had become accustomed to shopping in Yaohan as children. They gradually developed the habit of going to malls or department stores without intending to buy anything in particular. In other words, for them 'going shopping' had become a major leisure activity. This is well illustrated by a 1991 survey on Hong Kong youth aged between 12 and 24 years. Asked about participation in twenty activities, secondary school children and working youths of the same age listed shopping among their five most frequent activities (Chan and So 1992). This had not been the case in the 1970s (Department of Sociology and Social Work 1972).

## Conclusion

This chapter has explored the emerging cultural identity of Hong Kong people using the popularity of Yaohan in 1980s Hong Kong. We have demonstrated that Yaohan was a regional supermarket, adopting the business model of Japanese supermarket: locating stores close to residential areas and providing daily necessities and fresh food for ordinary customers. This business model made Yaohan's operation different from other retailers in Hong Kong. Yaohan in fact represented a new retail format. This new shopping format matched the cultural logic of the identity formation of *heung gong yahn*, they longed for something between Chinese/traditional/old and western/modern/new. Yaohan was thus welcomed by the Hong Kong customers, especially by the new middle-class families. Consequently, shopping in Yaohan became associated symbolically with the new middle class and Yaohan became a status symbol for them. Our contribution, therefore, lies in delineating the logic of the identity of Hong Kong people through studying the consumption of Yaohan by Hong Kong people.

There are several major implications to this. Firstly, scholars need to pay attention to the distinction between identities and presentations of self. Conflating the former with the latter runs the risk of turning people into postmodernists who all too easily swap their identities. It follows that better sense can be made of cultural identity through the study of consumption than through questionnaires or telephone surveys. Rather than asking the people of Hong Kong directly, we are convinced that studying their consumption patterns can tell us more about their identities. This active assemblage of a lifestyle by individuals or groups and how that lifestyle in turn shapes their identities has been documented by many other scholars (Miller 1987), and we consider this chapter to be part of this intellectual tradition.

This further points to a general anthropological observation: the relationship between identity and its signifiers is culturally constituted and therefore arbitrary – in the sense that there is no necessary relationship between identity and signifiers. The Chinese national flag does not necessarily imply a Chinese identity, while the identity of Hong Kong people can be symbolized by shopping at a *Japanese* supermarket. Human acts are culturally motivated, as we have seen; the cultural reason that Yaohan can serve as the marker of the identity of Hong Kong people was not only due to its *Japanese* background or its non-eastern and non-western image, but also to the cultural significance attached to the image by the logic of its identity of ***heung gong yahn***. Perhaps this is why the logic of the formation of Hong Kong identity is important, a crucial point that unfortunately has been neglected in previous studies.

Finally, it is clear that Yaohan's venture into Hong Kong involved different structures with different cultural logics stemming from both Japan and Hong Kong. These structures reciprocally mediated with one another to produce a historical effect that could not be predicted from any one of these structures, as seen in Yaohan's success in the late 1980s. All of these mediations are what Sahlins calls 'a work of cultural signification, which can be similarly described as the appropriation of local phenomena that have their own reasons in and as an existing cultural-historical scheme' (2000a: 301).

The result of such cultural signification is that '[t]he historical significance of a given incident –its determinations and effects as 'event' – depends on the cultural context' (ibid.). Yaohan's expansion in Hong Kong had its own rationale, including for example, Wada's idea that Yaohan should retain its own identity. However, the cultural effects of its expansion into Hong Kong depended on the socio-cultural context there. The initial success of Yaohan points to the fact that the form and extent of the cultural effects of Yaohan cannot be gauged simply from the 'objective properties' of Yaohan's Hong Kong venture, rather they hinged on the way that those properties—for example, Yaohan's adoption of the business model of Japanese supermarket—were symbolically mediated by the emergence of Hong Kong's new middle class, and the identity shift among the new middle class in the 1980s. In other words, the concept of homogenization is not enough of an explanation for the social effect of cross-cultural migration of cultural goods.

The objective properties of Yaohan's venture must not be treated as irrelevant, either. Yaohan's adoption of the Japanese supermarket business model made its operation in Hong Kong distinctive, which further made a difference because Yaohan represented a new retail format that matched the cultural logic of Hong Kong's new middle class's identity formation. Consequently there is no simple creolization explanation, either. The crucial aspect for us to consider as theorists is that both the circumstances of the historical conjuncture and the articulation that take place between the local and the global matter. In short, globalization is a complex process which should not be forced into singular theoretical frameworks.

## References

Babbie, E. (1983). *The Practice of Social Research*. Belmont, CA: Wadsworth.

Barth, F. (1969). *Ethnic Groups and Boundaries: The Social Organization of Culture Difference*. Prospect Heights, IL: Waveland Press.

Becker, H. (1956). 'Interviewing Medical Students'. *American Journal of Sociology* 62(2): 199–201.

Chan, S. C. (2010). 'Food, Memories and Identities in Hong Kong'. *Identities* 17(2/3): 204–27.

Chan, M. J. and C. Y. K. So (1992). *Mass Media and Youth in Hong Kong: A Study of Media Use, Youth Archetype and Media Influence*. Hong Kong: Commission on Youth.

Chen, K. H.Y. (2008). 'Positioning and Repositioning: Linguistic Practices and Identity Negotiation of Overseas Returning Bilinguals in Hong Kong'. *Multilingual* 27: 57–75.

Chiu, S. W. K., K. C. Ho and T. L. Lui (1997). *City States in the Global Economy: Industrial Restructuring in Hong Kong and Singapore*. Oxford: Westview Press.

Chun, A. (1996a). 'Discourses of Identity in the Changing Spaces of Public Culture in Taiwan, Hong Kong, and Singapore'. *Theory, Culture & Society* 13(1): 51–75.

—— (1996b). 'Fuck Chineseness: On the Ambiguities of Ethnicity as Culture as Identity'. *Boundary 2* 23(2): 111–38.

—— (2009). 'On the Geopolitics of Identity'. *Anthropological Theory* 9(3): 331–49.

—— (2010) 'Hong Kong 'Identity' after the End of History'. In Y.W. Chu and E. Kit-wah Man (eds), *Contemporary Asian Modernities*, pp. 167–90. New York: Peter Lang.

Comaroff, John and J. Comaroff (1992). *Ethnography and Historical Imagination*. Boulder, CO: Westview Press.

Department of Sociology and Social Work (1972). *A Study of the Leisure Activities of Youth Labourers in Hong Kong*. Hong Kong: Hong Kong Baptist College.

Evans, G. and M. S. M. Tam (1997). 'Introduction: The Anthropology of Contemporary Hong Kong Identity'. In G. Evans and M. S. M. Tam (eds), *Hong Kong: The Anthropology of a Chinese Metropolis*, pp. 1–24. Richmond, Surrey: Curzon Press.

Fung, A. (2001). 'What Makes the Local? A Brief Consideration of the Rejuvenation of Hong Kong Identity'. *Cultural Studies* 15(3/4): 591–601.

—— (2004). 'Postcolonial Hong Kong Identity: Hybridising the Local and the National'. *Social Identities* 10(3), 399–414.

Goffman, E. (1959). *The Presentation of Self in Everyday Life*. New York: Anchor Books.

Hall, S. (1994). 'Cultural Identity and Diaspora'. In P. Williams and L. Chrisman (eds), *Colonial Discourse and Post-Colonial Theory: A Reader*, pp. 392–403. New York: Columbia University Press.

Itagaki, H. (1990). *Yaohan*. Tokyo: Pāru.

Kim, J. and S. H. Ng (2008). 'Perceptions of Social Changes and Social Identity: Study focusing on Hong Kong Society after Reunification'. *Asian Journal of Social Psychology* 11: 232–40.

Larke, R. (1994). *Japanese Retailing*. London: Routledge.

Leung, B. K.P. (1996). *Perspectives on Hong Kong Society*. Oxford: Oxford University Press.

Lui, T. and T. W. P. Wong (1992). 'Reinstating Class: A Structural and Developmental Study of Hong Kong Society'. *Social Sciences Research Centre Occasional Paper 10*. Hong Kong: Department of Sociology, the University of Hong Kong.

Ma, E. K. and A.Y. H. Fung (1999). 'Resinicization, Nationalism and the Hong Kong Identity'. In C. So and J. Chan(eds), *Press and Politics in Hong Kong: Case studies from 1967 to 1997*, pp. 497–528. Hong Kong: Hong Kong Institute for Asia-Pacific Studies.

—— (2007). 'Negotiating Local and National Identifications: Hong Kong Identity Surveys 1996–2006'. *Asian Journal of Communication* 17(2): 172–85.

Miller, D. (1987). *Material Culture and Mass Consumption*. Oxford: Blackwell's.

*Nikkei Ryūtsū Shinbun*. (1993). *Ryūtsū gendaishi*. Tokyo: *Nihon Keizai Shinbun*.

Okano, M. and H. W. Wong (2004). 'Hong Kong's Guided Tour: Contexts of Tourism Image Construction before 1997'. *Taiwan Journal of Anthropology* 2(2): 115–53.

Sahlins, M. (1999). 'Two or Three Things that I Know about Culture'. *JRAI* 5(3): 399–421.

—— (2000a). 'The Return of the Event, Again'. In his *Culture in Practice: Selected Essays*, pp. 293–351. New York: Zone Press.

—— (2000b). 'Goodbye to *Tristes Tropes*: Ethnography in the Context of Modern World History'. In his *Culture in Practice: Selected Essays*, pp. 471–500. New York: Zone Press.

Sato, H. (1978). *Nihon no Ryūtsu Kikō*. Tokyo: Yūhikaku.

Sinn, E. (2009). 'In-Between Place: A New Approach for Hong Kong Studies'. In S. Wong and W. Chan (eds), *Rethinking Hong Kong: New Paradigms, New Perspectives*, pp. 245–304. Hong Kong: Centre of Asian Studies, The University of Hong Kong.

Tse, W. and K. Siu (2010). *Guomindang zhi xianggang bainian shilue* [The Brief History of Kuomintang in Hong Kong]. Hong Kong: 中華文教交流服務中心.

Tsuchiya, T. (1991). *Yaohan Wada Kazuo*, Tokyo: Kyōbunsha.

Wada, K. (1988). *Yaohan Inori to Ai no Shōnindō*. Tokyo Nihon Kyōbunsha.

—— (1992). *Yaohan's Global Strategy: The 21st Century is the Era of Asia. Hong Kong: Capital Communications Corporation Ltd.*

Wada, K. et al. (1994). *Yaohan Runessansu*, Tokyo: Kyōbunsha.

Wong, H. W. (1998). 'From Japanese Supermarket to Hong Kong Department Store'. In K. L. MacPherson (ed.), *Asian Department Stores*, pp. 253–81. Surrey: Curzon Press.

—— and H. Y. Yau (2015). 'There is No Simple Japanization, Creolization or Localization: Some Reflections on the Cross-Cultural Migration of Japanese Popular Culture to Hong Kong'. In H. W. Wong and K. Maegawa (eds), *Revisiting Colonial and Postcolonial: Anthropological Studies of the Cultural Interface*, pp. 39–68. Los Angeles, Calif.: Bridge 21 Publications.

Wong, Y. R. (2004). 'When East Meets West: Nation, Colony, Hong Kong Women's Subjectivities in Gender and China Development'. *Modern China* 30(2): 259–92.

DOLORES P. MARTINEZ

# Global Technologies, Local Interventions: Or Musings on Japanese Film

## The Opening Sequence

In post-war Japan and, to some extent, within Japanese studies, the dominant discourse has been one of incommensurable difference between Japan and the rest, leading to Orientalist and Occidentalist conceptualizations that East and West are dissimilar, as well as the idea that Japan as the locus of an eastern modernity is so distinct that it is unique (Befu 1992, 2001). In short the focus has been on Japanese disjunctures (pace Appardurai 1990): its different capitalism, business practices, sense of self, family ideals, educational values and even its popular culture. This chapter examines not disjunctures, but several conjunctures through an exploration of some issues relating to Japanese cinema. The argument is that global postmodernity and its discontents (Bauman 1997) have much in common, nationalistic discourses notwithstanding, while also noting that the phenomenon of media crossing borders – the subject of new studies in transnationalism (i.e. Iwabuchi 2015) – began at the moment new media technologies were invented.

To make the above point I use examples of 'national' filmic works and actors that have travelled and I rely on the concept of assemblage as discussed by Deleuze and Guattari (1986) in which the process of assembling can lead to the creation of series; and these can be anthropologically studied as social institutions as Latour argues (2007). However, Deleuze and Guattari consider another aspect of assemblage which seems often ignored: that dissembling, reassembling and deterritorializing also can result in the *creation* of something new. Thus the 'transnational' product, to use the current buzz word in media studies, stands not only as an example of western dominated global flows, neoliberal business deals, different nations' soft power or arguments over copyright (all issues included in Ong and Collier's 2005 volume), but also should be considered as the possible inspiration for localized artistic creativity.

To make this point, it is useful to consider Sahlins' (1981:35) concept of the 'structure of the conjuncture'. This structure is created through the encounters between different societies each of which have their own dynamics that 'meaningfully define the persons and the objects that are parties to it. And these contextual values, if unlike the definitions culturally presupposed, have the capacity then of working back on conventional values' (ibid.). For Sahlins understanding these conjectures explains how social reproduction takes place in the long run, a process in which an apparent social 'transformation' may well conceal unchanging attitudes, practices and concepts, while apparently static 'traditions' mask radical social change. For Sahlins (1981, 1985, 1995) the key examples, analysed repeatedly as part of his debate with Obyesekere (1992), are the 1778–9 arrival and death of Captain Cook in the Hawaiian Islands. This encounter led to profound changes *not only* for the Hawaiians *but also* for the Europeans who had imagined a Pacific colonialism that would be benevolent because of the islanders' 'peaceful' nature. With Cook's death this largely British political imaginary underwent a massive revision, although it did not stop their colonial expansion; perhaps this is why Sahlins makes rather less of European conceptual shift than of the devastating changes for the Hawaiians.

The parallels between the Hawaiian and Japanese encounters with the West are many, although they may seem allegorical. Briefly put, Japan's opening to the West after Admiral Perry's Black Ships visited in 1852, although also forced and ending in the nation's incorporation within global capitalism, has not resulted in the tragic ending of things Japanese in the way that the Hawaiian encounter could be said to have done with its almost complete decimation of the indigenous population and incorporation into the United States as a state. Instead it has led to a reification of what it is to be Japanese, *nihonjin*, in a reworking of the state's and individuals' identities that continues to this very day. The discourse of a forced westernization, the fear of losing 'traditional' values,[1] linked to the demise of a distinct way of life, is a powerful tool in the construction of Japan's modern identity (Yoshino 1992). Continuity and change are the dominant tropes used to describe the current situation. Japan became the globe's second largest economy, and now occupies third place; it has become one of the globe's most technologically advanced societies; and it is politically powerful within various international organisations – all while clinging to a notion of uniqueness, of difference, so total that it must 'remake' the foreign (Tobin 1992) in order to survive the onslaught of modernization. Conversely we need to understand that in its resistance to, incorporation of and involvement with the outside world, in its various conjunctures with the West, East Asia and modernity, Japan is taking part in transformative processes that are not unique and which end in both reassembling a quality that is seen to be Japanese (however difficult that may be to define) and in assembling what is now vaguely termed 'the global'.

That is not to say that the experience of globalization is exactly the same for any of the societies undergoing the process: there are always imbalances in power. This should not deter scholars from looking for more nuanced ways to portray global encounters than those dictated by the terms so often used: clash, shock, impact, homogenisation or hybridity. Moreover we need to be aware that despite any inequities, these encounters are never one way and, most importantly, they are rarely finite or conclusive.

To argue my case I use examples from Japan's long history of involvement in the global sphere of filmmaking. Here we need to note how discussions of national cinemas implicitly distinguish between technology, defined as 'machinery and equipment based' on the 'application of scientific knowledge for practical purposes', from technique: 'the way of carrying out a particular task, especially the execution of an artistic work' (Shorter OED).[2] In this case the technology is that of the moving picture camera,

1 Of course, the Japanese had to first decide on and sometimes invent the nation's traditions and this was not a simple endeavour, see Gluck (1987) and Vlastos (ed.) (1988).

2 It is difficult to find work on technique *and* technology in film theory – most of the written work is either practical or assumes a basic understanding of the technology

introduced to Japan in the nineteenth century, not long after its invention, and the technique, somewhat more ineffable, is that execution of an artistic work that produces a *Japanese* film. My argument is that the focus on the Japaneseness of the Japanese film hides a simple fact: that the world of cinema is global; it is the result of countless interactions at many levels, leading to changes at home (wherever that might be) and abroad (however that is defined); and that Japan has played a key role in transforming not just its own cinema, but that of others as part of this constant assemblage and reassemblage of images, techniques and technology.

## On a Global Presence

In 2003 the Japanese film director, actor and comedian Takeshi Kitano won the Silver Lion special Jury Award at the Venice Film Festival for his film *Zatōichi* (2003) and thanked 'Kurosawa *sensei*' in his speech. Fifty-two years after Kurosawa's *Rashōmon* had taken the Golden Lion, and five years after his death, this moment was seen to be a validation of Kurosawa's great talent by his fans. There were various ironies about this: Kurosawa had fallen out with much of the Japanese film industry in the 1960s and although he had continued to win awards in his native country throughout his career, critics had tended to say that 'he was not Japanese enough'. Five years after his death, however, he was being re-instated as a great Japanese filmmaker: the newly revamped National Film Centre in Tokyo had shown a season of his films early in 2003 and has continued to feature Kurosawa's works in other festivals. Kitano's public adulation apparently marked a shift in the Japanese public's perception of Kurosawa's work: perhaps he was finally Japanese enough. However, as I learned during a research year in Japan in 2003–4, many young Japanese have never seen his work.

while discussing techniques. However, a recent collection entitled *Techniques et Technologies du Cinéma* (Gaudreault and Lefebvre 2015) tackles this issue.

Kitano's statement in Venice also interested me because of an incident a few years previously: I had argued in a lecture that Kurosawa had influenced Kitano's work. One of my students had promptly taken the opportunity, while being one of his minders at the London Film Festival, to ask Kitano about his relationship to Kurosawa. 'No', he is supposed to have replied, 'I see myself as being more like Clint Eastwood.' We discussed this in class, noting that Eastwood owed his early fame to starring in a remake of a Kurosawa classic and often used images and camera setups that invoked the master. Discussing this later with the Japanese film expert Donald Richie, he told me another ironic tale: Kazuko Kurosawa, the director's daughter and Kitano's costume designer for *Zatōichi*, had berated him for his public adulation of her father: 'He never used violence for violence's sake as you do', she was supposed to have said, referring to Kurosawa's reputation as a humanist filmmaker (Richie, personal communication).

This may explain why, in interviews, Kitano has noted that Kazuko Kurosawa shrugged off his attempts to demonstrate similarities between his work and her father's. For example, Tom Mes, a journalist for the *Midnight Eye* asked:

> *You've talked about the Kurosawa influence on some of the scenes of Zatoichi, particularly the swordfight in the rain. Kurosawa's daughter Kazuko made the costumes for the film, so how did she feel about that?*
>
> KITANO: While we were filming the rain scene I asked her 'Don't you think it looks like one of your father's films?' and she replied: 'Not at all.' Then later with the scene of the retarded boy who wants to be a samurai, I asked 'Don't you think this is just like Dodes'kaden?' and again she said 'Not at all' (laughs). Those moments aren't really homages, they're more like winks, funny little references.
>
> (Mes 2003)

I make these points *not* to re-explore the issues of transcultural translation through the work of Kurosawa (see Martinez 2009),[3] but because these

3 While I examined the transcultural in detail in my 2009 *Remaking Kurosawa*, in this chapter I specifically look at examples of technology and technique knowledge transfer within the global context.

examples neatly encapsulate some of the themes that need to be understood in relation to Japan's film industry as both a local and an internationally consumed and constituted product. The Japanese film industry is not a hermetically sealed and hermeneutically inaccessible art form, but is enmeshed in a dynamic and mutually constitutive relationship with a global industry that constantly attempts to bridge any disjunctures created by disparities in money, technology, culture or even differences in political ideology.

Succinctly put: how is Japaneseness or non-Japaneseness recognized in a film? The eminent Japanese film critic, Yomota, argues that 'Japanese film' is an ambiguous concept given Japan's long history of making movies. This involvement encompasses: the colonial era with studios existing both at home and abroad; the existence of non-Japanese directors; and includes various shifts in types of films that have been popular with audiences. Moreover it took the French discovery of the 'Japanese' film for the Japanese to know that they had created such a product – before that, they had just made movies (Yomota 2000). Yet this difficult to pin down quality[4] of difference amongst national filmmaking styles is embraced by contemporary nation-states as part of their nation branding, resembling the case study outlined by Katsuno later in this volume.

At stake in this is a point relating to the adoption of technologies as opposed to techniques, a point that comes with a great deal of politicized baggage. In Tobin's terms, Japan remakes western – particularly US – imports and, earlier, Burch (1979: 90) also took this line by offering the model of acceptance, rejection and adaptation as a way of understanding Japanese film's relationship with western cinema. Both sorts of analyses place Japan in the role of the passive receiver of new ideas, gradually becoming more active as it incorporates things western by adapting or remaking them. Taussig (1991) evokes Walter Benjamin to critique such arguments by noting how

4 Choi demonstrates how difficult it is to define the term national cinema and argues that the concept works best as a sub-category that is 1) used as a label to differentiate a local film industry from Hollywood; 2) is used to designate a corpus of films by a group of filmmakers who share an aesthetic or ideological framework within a specific historical context; or 3) it may be associated with a 'new wave' of films that are seen as art films that have been produced within a nation-state (2006: 314–15).

they create alterity through the implication that *Others* (non-western, non-civilized) copy and remake, *we* originate. I argue that in fact Japanese filmmakers (and all filmmakers) dismantle the work of others, reassembling it in ways that are meaningful for their society. This deterriorialization is an operation that is 'a Process, one that is precisely interminable' (Deleuze and Guattari 1986: 48), but is *never* a passive act.

The technology of filmmaking arrived in Japan as a modern invention and so was value free, at least in terms of cultural context: a camera was a camera was a camera; nothing about it as a material object forced people to immediately confront issues of identity vis-à-vis the idea that they were 'copying'.[5] While some theorists (i.e. Carpenter 1976), argue that the addition of a new technology profoundly changes any society, the existence of typologies such as 'national film' contradict the view that these changes create homogeneity; on the contrary what we see is that importing new technologies opens up a space in which difference can be generated.

Technique cannot be entirely separated from technology: rather new objects offer possible actions (affordances) from which different people and societies may choose (see Gibson 1977). In the case of filmmaking such affordances lead to the frequent use of specific techniques. To repeat: there is nothing about the camera or the mechanics of editing to force one to do things in narrowly defined, homogenous ways. These techniques, or how the camera is used in different societies and in different eras, can reveal cultural processes of decision-making, political hierarchies, ways of seeing and representing the world that all are embedded in the diverse social practices through which a narrative is organized. Critics and audiences from outside the society may also assess the techniques used by a director (including long shots, stillness, close ups, etc.) to describe a film as typically Hollywood or Japanese or French, etc. Thus a film becomes an intervention: as a product born from the potential uses offered by the technology of the camera (and now the computer) and its culturally selected

5 See Nornes (2003: 1–47) for a description of how exciting urban Japanese found the concept of the moving camera and how some individuals quickly created their own home studios.

uses (techniques), it is situated between the audience's perceptions and the filmmakers' intentions. In such cases the film acts as an obstacle that can seduce us into looking at no more than the surface, into thinking of it as culturally bounded text which *must* be analysed within a national context.

Let us rewind and unravel some of the issues at stake by examining instances that span the era of filmmaking in Japan and elaborate on the issue of transferring techniques rather than technologies through a discussion of Sessue Hayakawa, the film festival circuit and building on the examples of Kurosawa and Kitano.

## International Superstardom

Best remembered for his role as Colonel Saito in *Bridge on the River Kwai* (Lean 1967), Sessue Hayakawa was born in Japan eight years before the first film was shown to Emperor Meiji on 15 February 1897. Hayakawa became a superstar in Hollywood silent films, an international heart-throb, whose salary was one of the highest in Hollywood in the decade spanning 1910 to 1920 (Miyao 2007). Anderson and Richie (1982) note another important fact: it was Hayakawa who, after the decline of his Hollywood career, returned to Japan and helped change the nascent film industry there; he is credited with introducing the notion that the camera could be moved. Burch (1979: 93) argues instead that the film production company Shochiku imported technicians from Hayakawa's own Hollywood-based production company and it was these Hollywood-trained technicians who revolutionized Japanese filmmaking, especially its film editing.

Hayakawa also was important in Europe. The film theorist Miyao notes that the French coined the term *photogénie* (photogenic) to describe how his acting technique made Hayakawa's screen presence so magnetic. Miyao notes: 'For them, the concept of *photogenie* (sic) was the basis of a new idea of film as a unique art form, thus Hayakawa of a new form of acting' (2007: 24). Within French theory the idea that cinema was an art form, capable of transporting its viewers beyond the real into a realm where people's souls and objects' essences could be perceived predated any North American ideas about film as art. It was French film theory, born in the constant re-viewings of Hayakawa's acting, which would eventually, through its concept of the auteur (the director as author), lead to the distinction between art house – often seen as national cinema – and mainstream or Hollywood cinema. Yet the foreign films perceived as an art house films in the West may be nothing more than mainstream films at home. What makes them seem radically different could be the language and setting, but difference can include types of camera shots, styles of acting, ways in which sound has been used or how the film was edited – all historically and culturally constructed forms of technique.

While new filmmaking techniques were imported Japan from Hollywood, they were in no way essentially American: the first narrative film, for example, may have been Porter's *The Great Train Robbery* (1903) made in New Jersey, but it inspired the Australian director Tait to make the first feature length narrative film, *The Story of the Kelly Gang* in 1906, each filmmaking society building on the back of the other. More importantly, the process of subtle editing was not part of the French Lumière brothers' work seen early on in Japan; early twentieth-century Hollywood developed its own style of editing, as did the French (calling it montage) and the Soviets – all of whom were studied by Japanese directors. So while the technology of filmmaking was brought to Japan from the West, this was followed by a process of skill acquisition that encompassed not just the Anglophone world, but also Europe (in particular Germany and Italy) and, increasingly into the 1920s and 1930s, the Soviet Union (Lannoy 2009).

This filmmaking 'toolbox' of techniques does not hold an unchanging set of tools. When Kurosawa rose to international fame in the 1950s, it was his use of three to five cameras that was considered ground-breaking by other filmmakers: Hollywood and the West tended to rely on a single, often stationary, camera except when filming musicals. The use of multiple cameras meant that Kurosawa was able to edit with a choice of shots made from many angles and copying this led to changes in Hollywood action films. The influence of this filmmaking technique on Hong Kong martial arts film was also immeasurable (see Yau 2010). Historians of Japanese film such as Yoshimoto (2000) have noted that, despite the West's admiration for these techniques as being Kurosawa's own, many of them were standard in Japan and used by other filmmakers, who were unknown or less famous in the West. Kurosawa also continued, as did other Japanese filmmakers, to use editing techniques developed in the era of silent film, long abandoned in Hollywood but not necessarily elsewhere. In contra-flow Kurosawa's 1960s graphic solution to the problem of realistically depicting a spurting artery (via a pump and chocolate sauce) in *Sanjūrō*, was to change European and US depictions of violent death. Cook (1998: 143–4) notes that in his attempt to replicate Kurosawa's realistic depiction of death by sword, the director Sam Peckinpah came up with the technique of using exploding squibs that showed an actor 'bleeding' after being shot.

More recently, Kitano's *Zatōichi*, combined computer effects for wounds and cuts to 'paint' in the blood (which is now common in western films), while relying on some Kurosawa camera and editing techniques to make the sword fights look fast-paced. This parallels the ways in which blockbusters (*The Last Samurai* [Zwick 2004], *The Hobbit* [Jackson 2012, 2013, 2014], *Mad Max, Fury Road* [Miller 2015][6] to list a few), have combined Kurosawa's technique of using several cameras to depict the sense of chaos from various angles when filming battle sequences, with a reliance on digital technology that can make one squadron look like ten; while Chinese films, such as Zhang's *Hero* (2002), use cheap extras to recreate Kurosawa-style battles.

The examples used here make a point about the constant movement of techniques as well as of new technology, or of novel ways of using old technology; a similar point could be made through the example of cartoons and anime. The pain-staking work by the hundreds of cartoon artists employed in Disney's early days eventually gave way to more mechanistic and less detailed drawing. Ghibli studio's anime challenged Disney and has led to a renaissance in the American studio, which had to raise its game to

6 This film also has a quick nod to Kurosawa's *Shichinin no Samurai* (1954); see if you can spot it.

compete with the Japanese, resulting in Disney's homage *Big Hero 6* (Hall and Williams 2014). Some of this has happened through the development of new technologies that come closer to the old hand drawn methods, while Pixar has led the way in completely computerized technologies, giving rise to new animation techniques. We can say that in both these cases, the flow was from the United States to Japan and back again, but this would ignore both societies' engagement with a global network in which the animation work might well be outsourced to the Philippines. Despite the rather blinkered assumption that American cultural imperialism is what Japan has had to struggle with and resist, the fact is that filmmakers watch each other's work all the time. New techniques and technologies move as fast as they are manufactured or seen on the screen. And if the new technology is not available to the ambitious filmmaker, they will, somehow, find a way to achieve the same effect with what old technology they have, often inventing a new technique along the way.

What can be said about these global processes is that until recently theorists have not seen them as interrelated series of changes – as the continuous ebb and flow of actual material objects across borders (film in its most basic form, cameras and other tools of cinematic technology) – but rather they have tended to be theorized as moments, abrupt – even violent – intrusions from one place to another. Thus this global movement has been conceptualized in terms of particular events and special people who come to be seen as nexus points. Film studies has long focused on movie stars, film directors or 'the story', for example, rather than the more prosaic to-ing and fro-ing that actually occurs – although the new wave of 'transnational' cinema studies is beginning to swell with a slightly different emphasis on the movement of cultural objects as globally marketed products. One way to understand this commercial aspect is to consider film festivals.

## International Awards

When *Rashōmon* won the Golden Lion in 1951, there were few film festivals and fewer awards for films than exist now. Critical acclaim in the art house sense was certainly associated with the post-war European international festivals, while more 'local', nationally-based awards were often self-congratulatory. As the business aspect of film distribution changed, post-1970s, and following the new wave of independent filmmaking in the 1990s, the international market for 'foreign' films has focused on these festivals. There are more than 100 such events worldwide, with Europe hosting the most, followed by East Asia and then Africa: festivals are where distributors go to find new material. They are the wholesale market from which the new director, star or genre is launched into the wider world of retail, which is fraught with issues of distribution rights, copyright infringement, and royalties (Ortner 2013). Moreover, the possibility for greater international recognition has to be balanced against the financial gamble that the director, producer, stars, camera crew, scriptwriters might be able to do it again. So festivals are not just where distributors vie for the right to make money by purchasing the rights to distribute 'foreign' films to new audiences, but also are where producers come to decide whether or not they are impressed enough by a film to support a director's future project. Film festivals are talent contests on a large scale with all the pitfalls, responsibilities and rewards that such contests involve.

As with all talent contests, we need to ask what is being judged and by whose standards. For example, it could be argued that it was a foreign gatekeeper's intervention that led to *Rashōmon*'s success. When selecting a film for Venice in 1951, Donald Richie chose what he thought was the best Japanese film of that year. This choice could be seen as being informed by Richie's understanding of film, which in turn was based on his western education and experience. Therefore the gatekeeper is the arbiter of quality and what, in the world of 'foreign' film, might succeed internationally. Yet, already in 1950, Richie was well-versed in most aspects of Japanese filmmaking and was a great admirer of Ozu rather than Kurosawa. However, when asked to choose a Japanese film to send to Venice, it seemed obvious to him that, in 1950–51, *Rashōmon* was the obvious choice (private communication). Objectively, it is clear that within the canon of Kurosawa's work, *Rashōmon* relies less on a 'western' grammar of filmmaking than some of his later films. In fact, it is not clear in which tradition of filmmaking it should be classified: it was, for many critics – Japanese and western alike – a film with a confusing narrative: it posed a mystery but did not solve it. That it looked beautiful and was from a poor country, a former enemy of the western Allies, was a surprise to many foreign critics (see Harrington 1987), while its visual techniques and strong sense of aesthetics seemed out of keeping with the post-war idea of a diminished Japan.

Aside from the business side of this moment, which brought a vibrant post-war industry to the attention of western filmmakers, producers and scholars, and resulted in a market for certain types of Japanese films; what is also significant about this event is the idea that a narrative technique, the so-called, 'rashomon technique', has made its way into the genres of, particularly, detective and crime films. Many audiences around the world now take for granted that a mystery might be solved, on-screen, after each character has told the story from their point of view; that the various versions tell fragments of a larger story and can be assembled into a complete solution to the 'whodunit'. While this is not true to the spirit of *Rashōmon*, it is another example of the film as intervention: how, as an object born from technology and technique, a movie's locus between the audience and filmmaker allows it to be re-appropriated, reassembled and re-membered in new ways. The use of multiple points of view in a narrative is not uniquely Japanese and existed before *Rashōmon*, but the 'technique' is not named after Charles Dickens, Bram Stoker or Ryūnosuke Akutagawa – all fiction writers who used it – or Welles, who certainly did it in *Citizen Kane* (1941), but after a film that made an unexpected impact in 1951.

For scholars of the Japanese film industry, *Rashōmon* may have been an unusual film, but its techniques, the crew and cast Kurosawa worked with, the stories the film was based on, are all part of processes that need to be understood within the history of that industry (Yoshimoto 2000); this contextualisation renders Kurosawa's work rather less ground-breaking. For the theorizing anthropologist, it is the Japanese film industry's, as well as Kurosawa's, long engagement with the films and literature of the world outside Japan, that must also be considered as parts of this process, as is the ways in which the technology and it attendant techniques were absorbed into this industry, sometimes to be used as needed, often to be emulated, copied, and sometimes to be rejected. The film world is never static, nor ever as parochial as is represented by nationalists or some theories of national cinema. Foreign audiences may notice stark contrasts when presented with something new, as in *Rashōmon*, but they quickly come to internalize the difference. They may expect always to find that difference in a foreign film (hence Kurosawa's samurai films tended to be more popular than his modern dramas in the United States), while at the same time accepting the incorporation of the new 'foreign' technique or technology – use of multiple cameras and a change in editing styles – as perfectly natural when encountered in their local films.

Film theorists and filmmakers, on the other hand, try to keep track of such connections, while film critics and festival judging panels are always looking for that seemingly inexplicable quality that makes a film worthy of a prize. As previously mentioned, battle scenes will lead to Kurosawa comparisons, whereas stillness in films will often bring to mind Ozu and an extreme form of anarchic violence, borrowed from cop films in the United States, has come to be seen as the norm in Kitano's work as well as in the work of Quentin Tarantino – however, what matters is whether these techniques are reassembled in such a way that they feel 'novel'.

Moreover Kitano, like Kurosawa, should not be understood without reference to his exposure to Japanese, East Asian and western cinemas. *Zatōichi* may owe something to Kurosawa, a lot to the original films on the blind swordsman, something to Hong Kong and maybe a bit to Bollywood as well as Eastwood. It is the coming together of various genres, their reassemblage, to create something original (Todorov 1990) that impresses audiences, prize givers, producers, distributors and other filmmakers alike.

Audiences may notice the new, without considering what went before or the technological processes involved in achieving it. The film for them becomes, as noted above, an intervention, but one upon which they also can project their own opinions, ideas and interpretations. Film critics and theorists, as another category of audience, are concerned both with the 'meaning' of the final product and with the processes that have gone into producing it. However, the focus on what the film is about – what it means – can result in a sort of blindness concerning the very materiality of film itself, as Deleuze notes:

> For theory too is something which is made, no less than its object... A theory of cinema is not 'about' cinema, but about the concepts that cinema gives rise to and which are themselves related to other concepts corresponding to other practices, the practice of concepts in general having no privilege over others, any more than one object has over others. It is at the level of the interference of many practices that things happen, beings, images, concepts, all the kinds of events. The theory of cinema does not bear on the cinema, but on the concepts of the cinema, which are no less practical, effective or existent than cinema itself.
>
> (1989: 268)

To theorize apropos an Orientalist western film theory, which is only interested in Kurosawa (Yamamoto 2000) or Kitano for their exoticism, is both correct in some of its assumptions about what audiences expect of foreign films, but also incorrect in postulating that the 'closed' world of the Japanese film industry, or any non-Hollywood film industry for that matter, is accessible to outsiders only in the most superficial sort of way through selective filmic experiences. All films are open to the audience as active readers, no matter their point of origin.

For people in the film industry, in contrast to the everyday audience, a film is not only a complete product but a material object with various levels they can easily peel away through their expertise. When filmmakers watch a film they immediately speculate on the type of camera and film stock used, on the cinematographer's choice of shots, the sound, whether the editing works in terms of the narrative, whether in post-production the colour values were maintained and so on. This is their method for understanding both the quality of the filmmaker's technique (including whether it is a standard, culturally contextual technique) and thus the merit of the film. Consequently while filmmakers can hold their own when it comes to theorizing on the meaning of films, they tend to prefer to discuss the techniques and technology necessary to make the visual narrative succeed. In short, to return to a metaphor used above, they often ask: has the director used the best tools (both technique and technology) for the job?

Conversely, the audience tends to ask: What was that about? Did I enjoy it? While pleasure is not absent from a filmmaker's judgement of a fellow craftsman's use of technology and technique, their enjoyment is often followed by the question: Could I do that?

For the filmmaker always looking for something new to try, technology and technique are not seen as separate processes. The general public, for whom 'foreign' normally equals 'difference', that something indefinable which they have experienced and which they might only partially attempt to explain in terms of filmic technique – the use of colour, flashbacks, the scenery, the special effects – is something they might sum up in terms of content saying 'it was a very Japanese film, it had samurai (or yakuza or dark-haired ghosts)'.

## Wrapping Up

I have been arguing that the film, as a finished product, should be seen as an intervention between the processes which produced it – the crews, the technology, the technique of a filmmaker, the society in which it was made, the historical moment it was made in and the global movement of other films – and the audience, which 'reads' the film in a very different way than specialists might. Perhaps it would be better to say: the discourse of the 'national' film is a sleight of hand, an entertaining magic trick that is able to distract the audience from appreciating film's universal nature and from seeing how it is a reassemblage of old and new, local and foreign even as they are managing to understand the story it tells.

To label a horror film, to give another pertinent instance, 'Japanese' or J-Horror is to build certain expectations in a western audience of difference. *Honogurai Mizu no Soko kara* (Dark Water, Nakata 2002), for example, was praised in the West for its ability to frighten with few special effects, thus creating a psychological terror. Its director, Nakata, however, acknowledged that some of the film's most subtle effects came from his affection for, and engagement with, the 1963 film by Wise called *The Haunting* (see Totaro 2000); just as Ozu always acknowledged that his 1953 *Tokyo Story* was a reworking of McCarey's 1937 *Make Way for Tomorrow* (Nolletti Jr. 1997); and Kurosawa always spoke of admiring Ford's cinematic 'grammar' (see Cardullo ed. 2008: 24–30).

Film theory needs to better cope with both the idea that film is a medium that is both local and can be international not just because it is enmeshed within a global economy, but because it can be meaningful across societies. While media or communication studies often engage with the business side of the industry, this is seen to be a separate disciplinary exercise than that of film studies' focus on the text. The new wave in the transnational film has begun to look at global connections, but tends to not explain *how* it is that film and television can be transnationally appreciated despite apparent cultural differences.

Where does the example of Japan leave us? The same argument could have been made for Hollywood accepting, rejecting and adapting the technologies and techniques from other societies that has been made for Japan. I began with the one and not the other because the example of Japan's long engagement with certain global processes, in this case, filmmaking, is one that could also lead the specialists of Japan to reconfigure the post-war discourse that sees Japan's modernity as a conundrum. Is it too western, has it lost its Japaneseness, if not, how has it maintained it?

As always with such loaded questions, the answer is both 'yes' and 'no' – all societies have had to deal with the technology of modernity, and all have developed their techniques for incorporating and using that technology: they deterritorialize, dismantle and reassemble. If modernity is a 'project' then it is not solely a western project and we need to better acknowledge that through a more nuanced use of Appadurai's 'global flows'. An anthropology that focuses mostly on local assemblages of global processes only observes part of the process, for such adaptations often feed back into the global and are once again reassembled there.

More importantly we need to understand that 'the global' is also an assemblage. That is: global culture, while perceived as somehow homogenous and predominantly western dominated may more accurately be described as 'the relations between general human development and a particular way of life, and between both and the works and practices of art and intelligence' (Williams 1976: 91). Global culture grows out of an infrastructure that is capitalistic (Jameson 1991, Cazdyn 2002), but the culture of late capitalism is now untethered from its western moorings, becoming increasingly fluid in its nature. Consider this while drinking a cup of tea (originally from China) with milk (an Indian addition) and sugar (a new world product) in a Starbucks (an American company modelled on European patterns of consumption), while perusing your texts (a form of communication more favoured by northern Europeans and East Asians than North Americans) on your mobile (invented in the United States but further developed by the Japanese, Danes and Finns), which was probably made in China. As for those who chat or tweet with various friends at once while sitting isolated in an office, does that make them individualistic Westerners or Orientals with a strong adherence to the larger group? Or is this a new way of being an individual that is developing as we type away?

Many modern patterns of consumption have been shaped by *global* technical innovations and economic interactions. The resulting changes in our manners and modes of communication may be coloured by our local cultures (do not try having a long conversation on your mobile on a Japanese train, but feel free to ring a friend and tell the story of your imminent business deal on a British one), yet we tend to ignore the profound fact that we are all taking part in a new global culture of instant and portable communication.

When Japan decided to modernize, it embarked on an exciting, ambitious and fascinating adventure that has had both dazzling successes and awful failures – this refers not just to the film industry. A better understanding of Japan's place in the global flow would be an understanding that does not rely on ideas of copying, remaking, acceptance and rejection or domination and subordination, but which would reflect in more detail on the concepts of participation, innovation and mutual incorporation. To build on Sahlins, conjunctures should be taken to be part of processes that end in transformations both locally and globally. Succinctly put: global technology and local technique (or objects that travel and supposedly local manifestations of aesthetics as in the case of film) cannot be separated. So while national ideologies demand that we see a Japanese film as full of ineffable Japaneseness, this is a very much a discourse rather than an incommensurate difference in filmic techniques. We should discuss degrees of difference – a Japanese close-up, for example, is much closer than a British one, but is still a close up – rather than differences in kind that make understanding impossible.

Japan's film industry may owe a debt to the West and to Hollywood, but Hollywood owes an equal debt to Japan as Clint Eastwood acknowledged in a pre-Oscar ceremony interview for this two 'Japanese' films:

> I haven't seen Milestone's *All Quiet on the Western Front* [1930] in many years, but it looked at the First World War from the German point of view and maybe there's a certain similarity to *Flags of Our Fathers* and *Letters from Iwo Jima*. I'm an aficionado, like everyone else, of Akira Kurosawa. His Samurai movie *Yojimbo* became *A Fistful of Dollars*. My admiration for him as a director led me into the career I had with Leone and beyond. Circumstances. The wheel goes around.
>
> (French 2007)

Indeed the wheel does, but when it comes to 'understanding' non-Hollywood films, film scholars in general tend to trace only the wheel's distinctive tracks, searching for signs of indigeneity, while ignoring the commonality of our shared modernity. I contend that while anthropologists of Japan should not ignore the discursive formulations that underpin nationalistic notions of difference and support the discourse of Japan's essential distinctiveness, we also should treat such discourses with caution and subject them to objective analyses that include the broader contexts of historical contingencies, economic connections, political ideologies and the flow of the technologies that make up modernity and make us all modern subjects.

What it means to be a modern subject, however, is not completely clear. It is a potential, not a given. The West is not 'turning Japanese' (or Chinese) nor are the Japanese becoming 'too western'. We both may be enmeshed within the structures of late capitalism, but we should not take the outcomes of this fact as given; to do so is to ignore the intricacies of the relationship between technology and technique; the processes of dismantling and reassembling; the desire to create and the complexity of lives lived between conjuncture and disjuncture.

## References

Appadurai, A. (1990). 'Disjuncture and Difference in the Global Cultural Economy'. *Theory, Culture* & Society 7, 295–310.

Anderson, J. L. and Richie, D. (1982). *The Japanese film: Art and Industry*, with a forward by Akira Kurosawa. Princeton, NJ: Princeton University Press.

Asad, T. (1993). *Genealogies of Religion – Discipline and Reasons of Power in Christianity and Islam*. Baltimore, MD: The Johns Hopkins University Press.

Bauman, Z. (1997). *Postmodernity and its Discontents*. Cambridge: Polity Press.

Befu, H. (1992). *Otherness of Japan: Historical and Cultural influences on Japanese Studies in Ten Countries*. München: Iudicium.

—— (2001). *Hegemony of Homogeneity: an Anthropological Analysis of Nihonjinron*. Melbourne: Trans Pacific Press.

Burch, N. (1979). *To the Distant Observer, Form and Meaning in the Japanese Cinema*. London: Scolar Press.

Cardullo, B. (2007). 'Interview with Akira Kurosawa, *Cinema/1963*', trans. Y. Kamil. In *Akira Kurosawa Interviews*, pp. 24–30. Jackson: University of Mississippi Press.

Carpenter, E. S. (1976). *Oh, What a Blow that Phantom Gave Me!* St. Albans, England: Paladin.

Cazdyn, E. (2002). *The Flash of Capital: Film and Geopolitics in Japan*. Durham, NC: Duke University Press.

Choi, J. (2006). 'National Cinema, the Very Idea'. In N. Carroll and J. Choi (eds), *Philosophy of Film and Motion Pictures, an Anthology*, pp. 310–20. Oxford: Blackwell.

Cook, D. A. (1999). 'Ballistic Ballectics: Styles of Violent Representation in *The Wild Bunch* and After'. In S. Prince (ed.), *Sam Peckinpah's The Wild Bunch*, pp. 130–54. Cambridge: Cambridge University Press Film Books.

Deleuze, G. (1989). *Cinema 2: The Time-Image*. Trans. Hugh Tomlinson and Robert Galeta. Minneapolis, MN: University of Minnesota Press.

Deleuze, G. and Guattari, F. (1986). *Kafka: Toward a Minor Literature*. Trans. Dana Polan. Minneapolis, MN: University of Minnesota Press.

—— (1987). *A Thousand Plateaus: Capitalism and Schizophrenia*. Trans. Brian Massumi. Minneapolis, MN: University of Minnesota Press.

French, P. (25 February 2007). '"I Figured I'd Retire Gradually, Just Ride off into the Sunset". An Interview with Clint Eastwood'. *The Observer* <http://www.theguardian.com/film/2007/feb/25/clinteastwood.oscars>.

Gaudreault, A. and M. Lefebvre (eds) (2015). *Techniques et Technologies du Cinéma. Modalités, Usages et Pratiques des Dispositifs Cinématographiques à travers l'Histoire*. Rennes: Presses Universitaires de Rennes.
Gibson, J.J. (1977). 'The Theory of Affordances'. In R. Shaw and J. Bransford (eds), *Perceiving, Acting, and Knowing*, pp. 67–82. Hillsdale, NJ: Lawrence Erlbaum.
Gluck, C. (1987) *Japan's Modern Myths: Ideology in the Late Meiji Period*. Princeton, NJ: Princeton University Press.
Hall, D. and C. Williams (2014). *Big Hero 6*. Walt Disney Animation Studios.
Harrington, C. (1987). 'Rashomon and the Japanese Cinema'. In D. Richie (ed.), *Focus on Rashomon*, pp. 141–4. New Brunswick, NJ: Rutgers University Press.
Iwabuchi, K. (2015). *Resilient Borders and Cultural Diversity: Internationalism, Brand Nationalism and Multiculturalism in Japan*. London: Lexington Books.
Jackson, P. (2012, 2013, 2014). *The Hobbit*. New Line Cinema.
Jameson, F. (1991). *Postmodernism, or the Cultural Logic of Late Capitalism*. Durham, NC: Duke University Press.
Kitano, T. (2003). *Zatōichi*. Asahi National Broadcasting Company.
Kurosawa, A. (1950). *Rashōmon*. Daiei Motion Picture Company.
—— (1954). *Shichinin no Samurai*. Toho Company.
—— (1962). *Sanjūrō*. Toho Company.
Lannoy, Y. (2009). *The View from the Bridge: The Cinemas of Kurosawa and Eisenstein between East and West*. PhD Thesis, Birkbeck, University of London.
Latour, B. (2007, new ed.). *Reassembling the Social: An Introduction to Actor-Network Theory*. Oxford: OUP.
Lean, D. (1957). *The Bridge on the River Kwai*. Colombia Pictures Corporation.
McCary, L. (1937). *Make Way for Tomorrow*. Paramount Studios.
Martinez, D. P. (2009). *Remaking Kurosawa; Translations and Permutations in Global Cinema*. New York: Palgrave.
Mes, T. (11/05/2003) 'Interview with Takeshi Kitano'. *Midnight Eye* <http://www.midnighteye.com/interviews/takeshi_kitano.shtml>.
Miller, G. (2015). *Mad Max: Fury Road*. Kennedy Miller Productions.
Miyao, D. (2007). *Sessue Hayakawa: Silent Cinema and Transnational Stardom*. Durham, NC: Duke University Press.
Nakata, H. (2002). *Honogurai Mizu no Soko kara*. Honogurai mizu no soko kara Seisaku Iinkai.
Nolletti Jr., A. (1997). 'Ozu's *Tokyo Story* and the 'Recasting' of McCarey's *Make Way for Tomorrow*'. In D. Desser (ed.), *Ozu's Tokyo Story*, pp. 25–52. Cambridge: Cambridge University Press.
Nornes, M. (2003). *Japanese Documentary Film: The Meiji Era through Hiroshima*. Minneapolis, MN: University of Minnesota Press.

Obeyesekere, G. (1992). *The Apotheosis of Captain Cook: European Mythmaking in the Pacific*. Princeton, NJ, Princeton University Press.
Ong, A. and S. J. Collier (eds) (2005) *Global Assemblages: Technology, Politics, and Ethics as Anthropological Problems*. Oxford: Blackwell Publishing.
Ortner, S. B. (2013). *Not Hollywood: Independent Film at the Twilight of the American Dream*. Durham, NC: Duke University Press.
Ozu, Y. (1953). *Tōkyō Monogatari*. Shōshiku Eiga.
Porter, Ewin (1903). *The Great Train Robbery*. Edwin Porter.
Sahlins, M. (1981). *Historical Metaphors and Mythical Realities: Structure in the Early History of the Sandwich Islands*. Ann Arbor: University of Michigan Press.
—— (1985a). *How 'Natives' Think: About Captain Cook, for Example*. Chicago: University of Chicago Press.
—— (1985b). *Islands of History*. Chicago, University of Chicago Press.
Tait, C. (1906). *The Story of the Kelly Gang*. William Gibson.
Taussig, M. T. (1991). *Mimesis and Alterity: a Particular History of the Senses*. London: Routledge.
Tobin, J. J. (1992). *Remade in Japan: Everyday Life and Consumer Taste in a Changing Society*. New Haven, CT: Yale University Press.
Todorov, T. (1990). 'The Origins of Genre'. Trans. C. Porter In *Genres in Discourse*, pp. 13–24. Cambridge: Cambridge University Press.
Totaro, D. (2000). 'The 'Ring' Master: Interview with Hideo Nakata'. *Offscreen* 4(3). <http://www.offscreen.com/view/hideo_nakata>.
Vlastos, S. (ed.) (1988). *Mirror of Modernity: Invented Traditions of Modern Japan*. Berkeley, CA: University of California Press.
Welles, O. (1941). *Citizen Kane*. Mercury Productions.
Williams, R. (1976). *Keywords: A Vocabulary of Culture and Society*. Fontana.
Wise, R. (1963). *The Haunting*. Argyle Enterprises.
Yau, K. S-T. (2010). *Japanese and Hong Kong Film Industries: Understanding the Origins of East Asian Film Networks*. London: Routledge.
Yomota, I. (2000). *Nihon Eigashi* 100-nen. Tokyo: Shūeisha.
Yoshimoto, M. (2000). *Kurosawa: Film Studies and Japanese Cinema*. Durham, NC: Duke University Press.
Yoshino, K. (1992). *Cultural Nationalism in Contemporary Japan: A Sociological Enquiry*. London: Routledge.
Zhang, Y. (2002). *Hero*. Sil-Metropole Organisation CFCC.
Zwick, E. (2003). *The Last Samurai*. Warner Brothers.

HIROFUMI KATSUNO

# Branding Humanoid Japan

## Asimo

In August 2003, Junichirō Koizumi made the first official visit by a Japanese Prime Minister to the Czech Republic. During his three-day trip, local newspapers constantly reported on the Japanese delegation. What the media remarked upon most was not the primary purpose of the visit, namely the politico-economic rapprochement between the two countries, but rather, the actions of a special goodwill ambassador who accompanied Japan's leader: Honda's Asimo. Prime Minister Koizumi introduced Asimo at a government dinner party, where the world's then most advanced bipedal robot walked around, gave a greeting in Czech and shook hands with the then Czech Prime Minister Vladimir Spidla.

In the three days' most symbolic moment, entitled 'Home Visit' by the Japanese media, the robot placed a floral tribute of chrysanthemums, Japan's national flower, at the foot of Karel Capek's statue in the Czech National Museum. Capek was the playwright who popularized the word robot. Capek's robot, a fictional bioengineered organic being created for forced labour, appeared in a 1921 Europe still scarred by the First World War and clearly symbolized fears of a technological dystopia. In contrast, Asimo – a real robot from the Far East, encased in a clean, white shell resembling a spacesuit – represented a future where robots and human beings harmoniously cooperated. Significantly, it was this robot that toasted the friendship between Japan and the Czech Republic.

For the Japanese government the robotics demonstration's official purpose was to display, on a highly visible stage, Japan's supremely advanced technology to the international mass media. The strategic use of the robot as spectacle in this diplomatic setting reveals both Japan's technological superiority and its attitude towards new technology as much as it publicizes how the humanoid robot, born in Eastern Europe and grown into an icon of modernity, has been best materialized by Japanese scientists and engineers.

Since the beginning of the twenty-first century humanoid robotics has increasingly become a productive arena for creating Japan's national identity. Taking Asimo's diplomatic voyage as a point of departure, this chapter focuses on this identity construction within the framework of nation branding, exploring how this political investment shapes the cultural dimensions of humanoid robotics in a highly nationalistic manner. This practice of nation branding, particularly through a continuous series of discursive and affective performances of robotic spectacle, turns humanoid robots into politically useful cultural content, anticipated as capable of contributing to, and promoting, Japan's national image as 'cool' as well as being the futuristic nation of robotics.

Japan has been 'the robot kingdom' (Schodt 1988) since it became a leading producer and consumer of industrial robots in the 1970s, but twenty-first-century Japan is increasingly defined not merely as an industrial dynamo, but also as the cosmopolitan architect of a future relationship between advanced technology, human beings, and the environment.

Through analysing robotic spectacles, I will examine how a complex network of social actors – including robotics scientists and engineers in university and corporate settings, national and local governments, political and economic resources, various arms of the mass media network, popular culture markets, and amateur technologists – enact the processes of Japan's nation branding. This forms a discourse related to the national policy shift away from the roboticization of Japan, which took place during the technological nation building of the post-war period, to the Japanization of the robot under the global politics of nation branding in the twenty-first century.

National identity in the era of globalization increasingly takes the form of brand nationalism (Iwabuchi 2007, 2010): the formation of national identity along cultural, rather than political or economic, boundaries. I examine this brand nationalism through data gathered during my ethnographic research, which was carried out in 2006, 2007 and 2011 to 2012 among the community of amateur robot-builders that has formed around the *Robo-One* competitions, one of the most popular and frequently held spectacular events of humanoid robots. Though amateur roboticists first appeared on the periphery of this technological arena in the early 2000s, the innovators and early adopters of this new technology have been integrated into Japan's brand nationalism. Within this arena, the Japanese are depicted as magical inventors and innovators as opposed to followers or replicators. Thus amateur robot-builders – promoted as culturally authentic, technologically savvy *Japanese* citizens – and their enchanting technology embody the national brand. This nation-branding activity articulates not only Japan's and the Japanese people's culture, nationality, and identity, but, in shaping the story of the robot kingdom, also brands humanoid robots themselves as quintessentially Japanese cultural commodities.

## The Great Robot Exhibition

The Great Robot Exhibition, held at Tokyo's National Museum of Nature and Science from 23 October 2007 through 27 January 2008, showcased over 100 robots of various shapes and sizes. Drawing more than 250,000 visitors, it was the largest national robotic event since the 2005 Aichi World Expo, receiving financial and organizational support from the Ministries of Education and Economy. Held in Tokyo's central metropolitan area, the Great Robot Exhibition was a national event, offering Japanese and foreign tourists alike the opportunity to become more familiar with robot capabilities and experience the development of Japanese robot culture.[1]

The selection of 'partner robots' – mostly humanoids – for display at the Great Robot Exhibition must be understood within the Japanese

1 Additionally, the Exhibition has spawned a number of smaller-scale exhibitions based on the same concept across Japan, enhancing the paradigm of current Japanese robot culture.

government's intention to support robot development as a key growth industry in the twenty-first century. Underlying this is the Japanese population's much-publicized ageing and the politico-economic motivation to create robots to compensate for the concomitant decline in the workforce. Japanese corporations, including NEC, Honda, Hitachi and Toyota, have also committed themselves to advancing robotics technology, with the ultimate goal of smoothly integrating robots into human society. Aiming to display Japan's technological competitiveness on the global stage, this exhibition simultaneously staked out a distinctly Japanese cultural space at the centre of the global robotics' universe. The exhibition inventively wove together past, present and future; the pre-modern, modern, and post-modern with traditional art, cutting-edge technology, and fantasy. At the heart of this narrative was the supposition that the humanoid robot is uniquely significant to Japan.

Consequently, the exhibition articulated forward-looking and dream-inspiring discourses that were enhanced by the affective influence of technological spectacle. In the first exhibit room were the robots of yesterday's future and their material legacies, ranging from the tin toy *Tetsuwan Atomu* (Astro Boy, the star of Japan's first animated television series) to plastic models inspired by *Kidō Senshi Gandamu* (Mobile Suit Gundam), a long-running, immensely popular animated television series featuring robot wars in space. Passing through this room, visitors reached the main exhibition space, where there was a display of the pre-modern mechanical dolls called *karakuri* dating from the Edo period (1603–1867). Directly facing the *karakuri* was an imposingly large sculpture designed by the renowned robot anime director Oshii. Beneath this fantastical figure, Toyota showcased the human-like capabilities of their latest partner robots.

Along the other walls of the exhibition space, a number of robots with functional capabilities were arranged, evoking an imagined science fiction future, anticipated as lying just around the corner. Most of these were developed in the 2000s with immanently practical ends in mind. These included robots designed for childcare (NEC's Papero), communication (Mitsubishi's Wakamaru), and security (Tmsuk's Tmsuk-4), as well as entertainment robots (Bandai's Doraemon and Toyota's humanoids). Adjoining this main space was a demonstration room, designed specifically for Honda's Asimo, where visitors were offered a practical visualization of a near future in which robots and humans co-exist.

In this presentation, humanizing machines' value went unquestioned as humanoid robotics were uncritically celebrated and invested with an aura of magical wonder. Here, the humanoid robot as spectacle was given the power to affect the viewers' subconscious from outside a strictly defined discursive reference system (see Massumi 2002). Nevertheless, this robotic spectacle does not act solely as affect; rather, its affective impact enhances the ways that it acts discursively.

The nationalistic tone of the exhibition's introduction, written by the museum's curator, Kazuyoshi Suzuki, reveals the exhibition's Japan-centric agenda. Suzuki's introduction, displayed at the exhibition venue and published in the exhibition catalogue, notes:

> People in Euro-American cultures do not associate organic images, like those of human beings and animals, with robots. Robots that communicate with human beings, like *Atomu*, are distinctive to Japanese culture. How has the climate and culture (*fūdo*) that spawned *Atomu* been shaped? The *karakuri* of the past are a prime source. The tradition of *karakuri*, which spanned several hundred years, has helped foster a *fūdo* in the Japanese mind that enables Japanese to feel affection for these human-like dolls. In addition, since Japan is endowed with a rich natural environment and four distinct seasons, its people have developed a unique sensitivity – even the chirping of insects is pleasant to our ears. The Japanese have an attitude toward and curiosity for nature distinct from that found in Euro-American society. Because of this, a number of Japanese *karakuri*, contemporary anime characters, and functional robots have been created in the likenesses of humans, animals, and even the natural environment.
>
> (my translation)

Instead of detailing each robot from a technological viewpoint, the exhibition presented the robots as various manifestations of Japanese robot culture, and was preoccupied with the questions of how and why Japan came to be the world's leading producer of humanoid robots.

Suzuki's narrative faithfully echoes the dominant popular discourse surrounding Japanese robotics, as shaped by the increasing visibility of humanoid robots since the early 2000s. He emphasizes the influence of two cultural products, post-war manga and the anime *Tetsuwan Atomu*, as well as the pre-modern and early modern *karakuri* on the development of Japanese robotics. He mythologizes *Atomu* as the ideal, definitive form resulting from a culturally-specific and exceptional technological trajectory and recontextualizes the antique *karakuri* as the techno-cultural origin of Japan's humanoid robots. As such, Suzuki's narrative employs the rhetoric of *nihonjinron* (cf. Befu 2001), of the sort influenced by Watsuji, the philosopher who explored Japanese cultural uniqueness in terms of environmental influences (see Martinez 2005). Similar explanations recur in, and are reproduced by, popular books and magazines as well as academic engineering journals, and have resulted in a nationwide revival of these cultural products.

When the media ran television programmes and articles featuring robotics engineers and researchers speaking of the imminent symbiosis between human beings and robots, *Atomu* was frequently mentioned and experts discussed the degree to which Japanese robotics had succeeded in realizing this benevolent, dream-inspiring robot.[2] At the same time, *karakuri* have been rediscovered through an increase in museum exhibits and hobby kits, while books and magazine articles dedicated to *karakuri* imply that the contemporary humanoid robot embodies a Japanese aesthetics originating in pre-modern automata (c.f. *MonoMagazine* 2006).

Such narratives often attribute the Japanese affinity for robots to indigenous animistic beliefs, which do not make clear distinctions between inanimate objects and organic beings. Animism is often said to be the fundamental principle of the Shinto; following this logic, a nation that practises animism is also capable of considering robots neutrally. According to this theory, the animistic Japanese have naturally embraced robots, which represent no inherent threat, while the Judeo-Christian worldview has obstructed western innovation in the field of humanoid robotics

2 Some critics (Sena 2001, Yonemura 2004) refer to the idealization of *Atomu* in Japanese robotics as '*Atomu* ideology'. Sena (2001) criticizes the Japanese mass media for hastily attributing Japanese robotics' prosperity to the popularity of the robotic genre (particularly *Tetsuwan Atomu*). In fact, the television series was developed to help promote Japan's then nascent nuclear industry.

technology. In western societies, it is implied, the Judeo-Christian dualism between human/non-human and good/evil has precluded the value-free acceptance of the non-human robot.

Such essentialist discourses oversimplify or even ignore the actual complexity of the processes through which concepts about, and images of, robots have proliferated and diversified in Japan. The strategic re-situating of *karakuri* as precursors to the contemporary context is cited as evidence for Japanese robophilia's cultural uniqueness, as though Japan was the only place where automata prospered in the pre-modern era. On the contrary, automata were widely diffused throughout the pre-modern world ranging from Asia to Europe (Nelson 2001, Segel 1995, Wood 2003).

Similarly, the *Atomu* example contrasts images of the benevolent Japanese robot with western (especially American) images of evil, destructive robots such as the Terminator (Cameron 1984). In his essays, the science fiction writer Isaac Asimov refers to this fear of humanoid robots as the 'Frankenstein complex' (Asimov 1978, also see Dinello 2005, Rutsky 1999). Though there is undoubtedly some truth to this generalization about western robot paranoia, Asimov's 'three laws of robotics' outlined in the early 1940s[3] have been strongly influential in the imagination of more benevolent western robots such as Robby – originally appearing in the American film *Forbidden Planet* in the 1950s and returning in the 1960s television series *Lost in Space* – and the comic duo of R2D2 and C-3P0 in *Star Wars*. More importantly, these robotic characters have been widely consumed outside the countries in which they were produced. While iconic robots from the United States have shaped a large market in Japan, Japanese robotic stories and products, from the classic *Astro Boy* to the renowned contemporary robot and cyborg anime such as Oshii's *Kōkaku Kidōtai* (Ghost in the Shell, 1995), have also been widely consumed in the United States.

3 First introduced in Asimov's 1942 short story 'Runaround', these oft-cited Laws are the following: 1) A robot may not injure a human being or, through inaction, allow a human being to come to harm; 2) A robot must obey orders given to it by human beings, except where such orders would conflict with the First Law; and 3) A robot must protect its own existence as long as such protection does not conflict with the First or Second Law.

Thus by its very nature the robot is a transnational figure, combining elements of reality, fiction and metaphor from across cultures. Given the vast number of metaphors and images that have accumulated around the concept of the robot globally since the early 1920s, it is too simplistic and arbitrary to draw a dichotomous line between Japan and the West as an imagined Other; or to attribute the origins of the contemporary robot to a particular country. Images and ideas about the robot have developed through an endless spiral of intertextual and intercultural negotiations within a global media culture.

What must be asked is why is Japanese robotics situated in opposition to the West and how is such a discourse possible?

## Japan as a 'Nation Brand'

Expanding networks of mobility, capital flow, media and migration have facilitated the territorial restructuring of the world economy. The unprecedented decentralization of socio-cultural and politico-economic production at subnational, supranational, and transnational levels has complicated the meaning of nationhood. Therefore the effective manipulation of images in the service of national interests is more critical than ever to local economies and to the formation of national identities. The term nation branding has been coined to describe the active construction of heritage narratives as well as the process of modernization undertaken by state and corporate actors on the basis of domestic and foreign concerns as well as economic and moral ideologies (Anholt 2003). Similar to commodity brands' production and advertisement, these invented narratives are highly selective; celebrating certain meanings and myths, while strategically ignoring others, in order to shape the common images and associations that help create the nation's 'imagined community'. Aronczyk notes: 'national interests are broadcast to audiences at large, complicating or overriding the narrowcasting of traditional state-to-state diplomacy. Nation branding therefore serves as a form of pre-emptive management and control, a national discourse for a global context' (2008: 44).

The concept of nation branding bears similarities to Nye's (1990) concept of soft power. In contrast with the coercive hard power of military or economic assets, soft power, the ability to shape the preferences of others through attraction, 'appeals to state crafters as better suited to the "public" or "popular" diplomacy requirements of nation-states in the contemporary context' (Aronczyk 2008: 44). Nevertheless, while Nye's original argument, popularized in the 1990s, takes soft cultural power to be a supplement to, not a replacement for, hard economic power, they have increasingly become two sides of the same coin in the global politics of nation branding. As more and more countries become involved in a fierce global competition for foreign investment and tourism, governments concern themselves with facilitating trade, improving industrial competitiveness and increasing geopolitical influence. Since a nation's formation of identity is increasingly linked to marketing strategies, the national brand is thereby materialized as a cultural form in the global market.

Iwabuchi (2007) has noted that the government's role in the politics of nation branding is not necessarily to be a leader in cultural and technological innovation, but to put a national seal on cultural practices and products, in the hope that they become synonymous with its nationality. Iwabuchi calls this mode of nationalistic practice 'brand nationalism', referring to the 'uncritical practical uses of media culture as resources for the enhancement of political and economic national interests, through the branding of national cultures' (2010: 90). This commercial nationalism is thus not an expression of culture's commoditization, but of the culturalization of commodities, wherein the potential diversity within the national is minimized through the dynamic interaction between nationalism and globalism. Thus, while its ultimate goal is the promotion of politico-economic interests by attracting foreign audiences, the internal process of nation branding strives to unite and mobilize domestic citizens (Anholt 2003). Nation branding simultaneously seeks to inform the international perceptions of a state while providing a lens through which the state's citizens might view themselves. Consequently, nation branding is a confluence of nationalism and globalism within in a planet-wide market, where identity, culture, political-economy, business and nationality all come into play.

Therefore, the essentialist discourses that surround Japanese humanoid robotics, including the *Atomu* and *karakuri* narratives discussed above, have been shaped in close relation to a change in policy following a shift in national ideology. During the technological nation-building of the post-war period, state and industrial actors focused on fully robotizing Japan. Given the twenty-first century global politics of nation branding, these same actors now aspire to fully Japanize the robot. By claiming the cultural authenticity of this invented tradition of humanoid robots, they are attempting to establish or renew Japan's national image as the source of future robotics technology. Thus, arguments positioning *karakuri* and *Atomu* as scions of an essentialist Japanese technological line are part of a historical conflict between culture and technology in Japan that dates to the Meiji Era.

The struggle to reconcile these two supposedly oppositional forces was expressed in the Meiji slogan *wakon yōsai* (Japanese spirit, western technology), an expression of strategic hybridism in response to the challenges western cultural and technological influence posed. Through this practice, 'the putative Japanese national essence is imagined in terms of its exceptional capacity for [the] cultural absorption of the foreign' (Iwabuchi 2002:19). However, the narrative that animated the Great Robot Exhibition – the Japanization of the robot – is more aggressive and confident, for it was predicated upon a synthesis of Japanese culture and technology, which we might call *wakon wasai* (Japanese spirit, Japanese technology) by completely obscuring the facts in the history of robotics' development.

Given the Euro-American historical importance in this technological arena, how is such a discourse possible? To understand this, we must first consider the radical changes in Japan's scientific, technological and politico-economic climates that were affected by the transition between the pre-bubble to the post-bubble economy period in the 1990s.

## From Nation Building to Nation Branding

Japan's post-war history began with the process of technological nation building; thus the production and consumption of technological products have become deeply linked with being Japanese (Low 2003, Yoshimi 1999). Japanese technology's and business models' success 'catalysed the rise of a renewed sense of pride and optimism grounded in the success of a "Japanese-style economy" (Iida 2002: 164). The Japanese economy's remarkable growth sparked the techno-nationalistic discourse of Japanese uniqueness, contributing to the *nihonjinron* literature of the 1970s and 1980s. *Nihonjinron* theorists were part of a neo-nationalistic search for a new Japanese cultural identity (cf. Yoshino 1992). They argued that Japan's status as the first non-western nation to achieve technological excellence and mass industrialization demonstrated its cultural uniqueness.

It was not only nationalistic Japanese who bought into the ideas of *nihonjinron*. For instance, Vogel's *Japan as Number One* (1979) reveals a western appreciation of the Japanese socio-cultural system that allegedly enabled Japan's economic success. Here, another variation on the *wakon yōsai* terminology came into being, with Vogel arguing that America should embrace a policy of *yōkon wasai* (western spirit, Japanese technology). Thus foreign observers ostensibly validated *nihonjinron* discourses, equating Japaneseness with advanced technology during this high-growth period.

The association of Japanese culture with technology concomitantly appeared in popular cultural texts. The first connection may have been made in 1980s American cyberpunk fiction, such as Gibson's *Neuromancer* (1984), which associated Japanese culture with the technological future. However, as Japanese technology infiltrated the American political and cultural landscapes, a reactionary discourse developed, bred by panic and tinged with antagonism. The US media developed a particular image of Japan as the negative incarnation of hypermodernity, clearly seen in Crichton's 1992 novel *Rising Sun* and the 1993 film of the same name. In this western imaginary, which Morley and Robins (1995) call 'techno-orientalism', Japan appears simultaneously as a technological utopia and as a cautionary metaphor for the dangers of hyper-modernism. That is, 'Japan not only is located geographically, but also is projected chronologically' (Ueno 2002: 216), and its highly advanced technology is seen to lead to human alienation.

This nation that excelled at producing industrial robots was imagined as being populated by robotic people, emotionless and homogenous – an obvious development of older Orientalist discourses. In these media representations, the West retained its cultural integrity in the face of increasing technological development, while Japan lost its humanity and soul to technology. The symbolic association of robots with Japan was a complicated convergence of Japanese nationalist rhetoric and American representations of technological Japan. The relative congruence between *nihonjinron* portrayals of Japan by techno-orientalists and theorists was the degree to which they both placed technological superiority at the centre of Japan's cultural identity.[4]

Such nationalist discourses receded in the aftermath of the economic bubble's collapse, which was compounded by the acceleration of the globalization of the United States and the rise of new East Asian economic powers, notably China and South Korea. In Japan there was a sense of crisis over a perceived loss of scientific, technological and industrial competitiveness. This sense of crisis was felt in the robot industry as well. In 2012 Japanese industrial robots still accounted for approximately fifty percent of all industrial robots worldwide, which indicates Japan's continued dominance over the global politics of robotics technology, a position it has held since the late 1970s. Nevertheless, the industrial community in Japan became worried that the robotics technology in industrial manufacturing had reached a plateau given that the basic technology of industrial robots was developed between the late 1960s and the early 1980s.

4 In a similar vein, Kumiko Sato (2004) astutely analyzes how Japanese cyberpunk triggered the *nihonjinron*-driven discourse of the equation between Japanese culture and future technology by actively appropriating perceptions of Japan from American cyberpunk.

This sense of crisis led Japan to enact *Kagakugijutsu Kihonhō* (The Science and Technology Basic Law) in November 1995, which mandated that the government increase its spending on research and development and institute reforms in science and education.[5] In conjunction with the Basic Law, the Japanese government initiated a legislative process to enact a pro-patent policy in 1996. This ended with the execution of *Chitekizaisan Suishin Keikaku* (The Intellectual Property Strategic Program) in 2002 and the formation of an Intellectual Property Strategy Headquarters in 2003. Both actions were carried out under former prime minister Koizumi Junichirō's *kōzō kaikaku* (structural reform) programme and indicate a shift in national policy from the technological nation building of the twentieth century to 'making Japan an intellectual property-based nation' in the twenty-first century.[6]

The 2002 version of the Strategic Program cited the often-praised Japanese economic model as a major factor in the decline in Japanese industry's international competitiveness. It states that 'Japan has been content with its old-style industrial system as a result of past successes and has failed to drastically reform the conventional Japanese model amid the rapid changes in the environment in recent years' (Intellectual Property Strategic Program 2002). The Strategic Program emphasized the necessity of promoting a new knowledge-based economy as a means of stimulating and revitalizing original innovations for economic growth.

To this end, the programme set the goal of making Japan an intellectual property-based nation, which means expressly establishing a national direction emphasizing invention, in which the production of intangible assets is recognized as the foundation of industry. This national brand would be known for its production of 'information of value' including various technologies, designs, and brands as well as the content of music, movies,

5 This sense of crisis pervaded in millennial Japan. The 2006 version of Japan's *Kagakugijutsu Kihon Keikaku* opens with: 'it is never easy for Japan, a resource-poor country, to occupy an honorable position in Humanity. In fact, the country's future prosperity depends on the development of unique, outstanding science and technology.'

6 This excerpt is taken from the Intellectual Property Strategic Program (2006: 27).

and the like. This is a national policy underpinned by a vision of revitalizing the Japanese economy and society (Intellectual Property Strategic Program 2002). Thus, Japanese state actors are keen to rebuild the nation on the basis of intellectual property, demonstrating their awareness of the global politics of nation branding in the post-industrial world.

## Nation Branding Through Humanoid Robots

So it was that Honda's bipedal robot P2 – renamed Asimo – appeared amid the heightened anxiety of Japan's 'lost decade'. Honda's achievement in creating the world's first bipedal robot with a smooth walking motion renewed hope in the field of Japanese technology. It was believed that this invention might result in a profusion of spin-off technologies, especially in the fields of human-interface, network and software development, in which Japan lagged behind the United States. This would trigger an economic ripple effect by producing a new market for lifestyle support robots for Japan and other fast-ageing countries.

In conjunction with the Basic Science Law, the Japanese government has sponsored a number of conferences and workshops in order to identify and promote new ways to use robotic technologies in human society. Through the slogan *kyōsei* (co-living) with the robot, the Japanese government has re-encouraged and reinforced cooperation between industry and academia, launching next-generation robot projects that centre on the development of humanoid robots as welcome sources of social interaction and emotional exchange. This trend became a national phenomenon with the Aichi World Exposition of 2005.[7]

7 As bipedal technology became a reality with Asimo, Honda came to propose its own vision of the robotic future: at the Great Robot Exhibition, its display presented Asimo as a robotic domestic servant that would do grocery shopping and play with children.

This represents a marked difference from the philosophy of the robot kingdom that had existed up to the 1980s, which advocated robots as industrial tools with rational ends. The emergent discourse in the twenty-first century is characterized by a more human element, as evident from the use of the word *kyōsei*. In this vision, robots are expected to fill new roles as companions, caretakers and mediators between humans and the increasingly complex socio-technical environments we live in. *Wakon wasai* in the twenty-first century has taken on an almost messianic tone, repositioning Japan as number one, not in the business or industrial realms, but in addressing humanitarian needs by developing humane robots.

This difference is clearly articulated in a passage intended as a sort of conclusion to the Great Robot Exhibition, in which the curator Suzuki writes:

> I believe that the future of nature, humanity, and the robot will not be shaped in a better manner without such unconscious communication. It is a view of the robot that only Japanese culture holds…To hold 'The Great Robot Exhibition' in Japan, which is recognized as the leading country for robotics in the world, is significant for us – the field of robotics has developed as part of Japanese culture in a trajectory from the past, through the present, to the future. Going beyond robotics, we must consider the role that Japan should play in promoting the harmonious coexistence of humanity and technology, humanity and nature, and humanity and the earth. (my translation)

While 1980s *nihonjinron* literature primarily argued that Japan's cultural exceptionalism had facilitated its high-growth economy and advanced technology, the discourse surrounding current humanoid robotics goes even further, advocating a cosmopolitan, almost evangelical mission to chart humanity's future, technology and the environment through *wakon wasai*, a synthesis of Japanese aesthetics and technology.

Also notable is that this image of Japan in the era of nation branding is essentially different than the robotic Japan imagined in the Euro-American media in the period from the late 1970s to the early 1990s. Whereas that techno-orientalism was based upon a western antagonism toward Japanese economic power, today the overseas appeal of Japanese popular culture in such areas as anime, fashion, food, and consumer electronics has generated soft power for the nation. The contemporary Euro-American popular discourse regarding Japan is still undoubtedly marked by traces of techno-orientalism, although with a more positive, 'cool' spin, but this vision is no longer exclusively a western creation. To establish the image of Japan as the centre of the global map of technological innovation, the mechanism of nation branding today makes use of the very techno-orientalist narratives that were formerly used against it.

At stake here is not whether Japan is actually a cosmopolitan nation, or even the truth-value or sincerity of Suzuki's words. What I am interested in is the ways in which a series of next generation robots, displayed and made into spectacles in various kinds of public events, are used to generate an image of a cosmopolitan Japan based upon an optimistic techno-utopianism, a Japan that can solve global problems through state-of-the-art technology.

This is clearly represented by the word *yume* (dream) as the preferred trope to describe the field of humanoid robotics, as in the phrase *yume no gijutsu* (dream technology). In this dream-centred discourse, the question 'why humanoid?' is never asked by mass media. The dream of human-shaped machines becomes a magic spell, enchanting society with the public imagination of a bright tomorrow. However, contrary to popular expectations, outside of specially conditioned laboratory environments, most humanoid robots today can only perform as the high tech puppets and spectacular Others that entertain audiences from afar at robotics exhibitions.[8]

Certain engineers and scholars have paid judicious attention to this uncritical celebration of humanoid robots' emergence. For instance, Shigeo Hirose, a professor of engineering at Tokyo Institute of Technology, famous for his creation of mine defusing and rescue robots, criticizes the humanoid-centred approaches that fixate on the human-shaped designs which

8 As of 2013, the humanoid boom is gradually passing. This is partly due to the increasing difficulty in funding research during the global economic crisis in conjunction with the chaos following the 2011 earthquake. But, a more fundamental reason is that the humanoid technology shows a sign of leveling off as the same robots have repeatedly been displayed at expositions in the past years. Some researchers have told me that it will take another fifty years before humanoids will have practical uses.

limit the potential in robot development's design and utility (Kamoshida 2005). In comparison with bipedal movement, wheel-based mobility is much more stable and thus appropriate for high speeds. Also, there is no reason that robots need complex human hands to perform routine jobs. Similarly, the prominent Japanese roboticist Michitaka Hirose points out that 'robotizing our technological environments requires much less time and financial investment than domesticating general-purpose, human-shaped robots' (2007: 19). Additionally, Arai (2001), a Senior Researcher at the Intelligent Systems Research Institute at the National Institute of Advanced Industrial Science and Technology, argues that humanoid research does not reflect technological necessity, but the personal motivations and obsessions of researchers with creating human-shaped robots, which they tie to the futuristic visions fabricated by the mass media.

These critiques of the unquestioning celebration of the humanoid seem to be come from a stereotypical engineering perspective; engineers are expected to rationally and practically apply scientific knowledge in order to improve human life. At the same time it is worth asking why humanoid robots encounter such a modern pragmatism, as if the simple fact that they have been dreamed about justifies their creation. Regarding this contradiction, Matsubara[9] told me:

> Many scholars are somewhat confused with the current situation in which their research came to be spotlighted. Before the humanoid boom happened, they had worked in isolated ivory towers. However, once taken to the public front stage, we were required to make our research visible and understandable to the general public at large.
>
> (personal communication)

Another university roboticist told me that humanoid research was relatively easy to fund, as it appears both comprehensible and desirable to non-specialists. Therefore, the field of humanoid robotics' emergence results from a complex intertwining of individual desires and institutional structures.

9 Matsubara is a prominent scholar in artificial intelligence in Japan, and is known as a founder of Robo Cup, the international robotic soccer competition among pedagogical institutions.

The general public's obsession with mechanically recreating the human form may even surpass that of the academic and industrial researchers themselves; this inevitably affects funding in the academy just as much as in the state and in industry.

These are potential counter-narratives about the effectiveness of branding Japan as the 'happy robot' nation. But, they have never come to the fore vis-à-vis the powerful 'techno-romanticism' (Coyne 2001) that fantasizes and embraces the humanoid robot as a life-transforming technology, a key element to an idealized, utopian future. In this ever-increasing society of the spectacle, governmental, scientific, and industrial forces work together to create a national consensus on the need to move toward a robotic future. These same forces simultaneously sell an image of Japan as the global leader in robotics to foreign audiences, and also imply that Japan's role in providing technology to solve universal problems should grant Japan a prominent position among the leading nations of the world.

## Nation Branding, Regional Branding and the Robotic Spectacle

As part of the branding process, state and local government institutions actively showcase individual practitioners at public events. On the national level, amateur robot builders have been invited to have their robots perform at such venues as the 2006 Aichi World Expo, the 2007–8 Great Robot Exhibition and the International Robot Exhibition in 2007, 2009, and 2011. These venues offer amateur builders opportunities for the recognition and validation of their labours of love. For instance, Yoshimura Koichi, winner of the second *Robo-One* in 2003, designed a robot that won the Ministry of Economy, Trade, and Industry's robot award in 2006. Yoshimura's robot became the prototype for Kondo Kagaku's best-selling KHR series robot kit, which has since sold more than 10,000 units to consumers, including electronic hobbyists and university laboratory researchers.

At the local level, exposure is much more extensive. Fukuoka city is one of the most active supporters of robotic spectacles, and amateur robot builders are invited to public events on a monthly basis. These are held as part of the local government's project to make robot production a major industry in the region.[10] Robosquare, a robot museum established and maintained by the Industry Promotion Department of Fukuoka city, plays a central role in the region's robot culture. Organized and managed by a committee comprised of executive officers from industry, academia, and governmental agencies in and around Fukuoka, Robosquare was established in 2002 with the primary aim of making Fukuoka the Silicon Valley of Robotics.[11]

Shinkawa Shinichi, the first president of Robosquare, told me:

> Robosquare was launched as an institution with a missionary role for the field of robotics ... our role is to produce a dream of a robotic future as a stimulus for technological development in this field. Indeed, it is not us but the robots that give such a dream to the public. Our task is then to explain the social significance of the robot to Japanese society, and possibly to the world. We need to give concrete images of a future with robots.

The dream-inspiring robots of which Shinkawa speaks are much like those exhibited at the Great Robot Exhibition – robots designed to be domestic partners helping humans in their everyday lives. Since its establishment, Robosquare has displayed representatives of twenty-five different robot 'species' making a grand total of over 100 robots.

Shinkawa further explained that:

> Today, Japanese society pays attention to robots, which are designed to achieve *kyōsei* with human beings. I want to make Fukuoka the most prominent city in the world in this respect. I want to make Fukuoka the central outlet of such a dream, the dream of human-robot symbiosis... Market survey representatives from foreign countries sometimes visit us and ask questions like, 'How does Robosquare make a profit?' We

10 Fukuoka city has been designated a special unregulated district intended to encourage experimentation on robot use in public space. Kyushu University and Waseda University have facilities in Robosquare, conducting regular robot experiments in Fukuoka's public spaces.

11 See <http://robosquare.city.fukuoka.lg.jp/english/index>.

> are not profit-seekers, but dream-makers for the people, especially children. We – as an administrative institution – do what the private sector cannot do. No technology develops without a dream of the future, particularly in this field.

As a means of inspiring dreams about this futuristic technology, the robot museum not only coordinates shows featuring established robot giants like Honda and Sony, but also actively helps local citizens realize their robot dreams by involving amateur robot builders in the public events held in Fukuoka city. For example, amateur builders gather biweekly at Robosquare, to engage in robot battles or robot soccer demonstrations. In addition, Robosquare has held the Humanoid Cup, a robotic battle and soccer competition, over thirty times since 2002. The competition has been held not only at Robosquare, but also at prominent public locales including subway stations, department stores, science museums, and theme parks. The Humanoid Cup, furthermore, has been officially affiliated with *Robo-One* since March 2007, and now attracts amateur robot builders from all over Japan.

The general audience for these public robotic events lacks the technical expertise to fully comprehend the mechanics animating the spectacle. As such, robotic entertainment is all about the 'enchantment of technology' (Gell 1992), the affective fascination with the made-ness of things. This is a 'black-box' scenario in which the workings of objects are unavailable to those who see them. The robot has the capacity to transcend the audience's own embodied materiality, but the ways through which it can do so remain mysterious. This kind of relationship between human and technological object transforms human perceptual capacities in an almost magical way within the context of everyday life.

Shinkawa explained to me that he actively involved amateur builders in public events:

> Because they become models for the next generation of robo-philes. How many robots in the world are actually conducting smooth bipedal walking, running and even doing martial arts? In Japan such robots are made not only in university laboratories but also at home. The amateur builders symbolize Japan's robo-philia. Children may be inspired to join the field of robotics by interacting with the robot-builders,

and foreign tourists come to see the surprising power of Japanese technology with their own eyes.

Together with the city of Fukuoka, Robosquare is actively promoting a vision of *robotto ga aruku machi* (the town where robots walk) in an effort to establish the institution as the face of Fukuoka city robotics, and Fukuoka as the face of Japanese robotics, in hopes of attracting the attention of domestic and foreign tourists, as well as that of industry. Through such projects and spectacles, the robot is deployed in the regional branding of Fukuoka as part of the larger process of nation branding.

The media play an important role in the branding operation as well. The media coverage of minor robot-building celebrities is a given, but amateur robot builders who perform publicly also receive frequent media exposure in Japan.[12] These amateur roboticists almost never fail to mention the word *yume* when expressing their enthusiasm for robot building, and they often use this term interchangeably with the word *roman* (romanticism). In the predominantly male community of amateur robot builders, the connotation of *roman* is highly masculinized as it is particularly referred as *otoko no roman* ('male romanticism'). For these robot builders, the essence of *roman* in robot building is intimately connected with 'real engineering': that is, in the heart and soul of cutting edge technology (see Katsuno 2012). In other words, what lies behind contestants' passion for robot-building is neither profit-driven production nor theoretical engagement, but the romantic potential to materialize a childhood dream: the creation and possession of a real robot, the opportunity to give free reign to one's creativity and imagination through the design of a home-built humanoid.

Nevertheless, the pleasure in robot building is not simply a naive, male romanticized fantasy. Rather, it plays a powerful role in reconstructing the

12 During my fieldwork, I joined a Robo-One group as a robot builder, learning robot-making, joining an engineering community, and taking part in public robotic events. In my capacity as a neophyte robot builder, I was interviewed by national television stations, including NHK and TV Asahi; local news stations in Tokyo and Fukuoka; a local newspaper; a German national broadcast company; one robotics-related magazine; and two web-based magazines – all, surprisingly, within one year's time.

strong link between technology, masculinity, and the nation in the dream-centred discursive system of Japanese humanoid robotics. Humanoid robotics connects dreams at the personal level to those at the national level, as this popular cultural medium is simultaneously a productive site of individual desire as well as a representation of the collective dreams of society at large. As a result, the nation-branding operation legitimates this masculinized passion and dream as culturally Japanese, and celebrates these builders as patriots exemplifying the national brand.

As a result, robot-builders have become highly self-reflexive with regard to how they are involved in the formation of Japanese robot culture on a grand scale. For example Horikawa, a 40-year-old systems engineer for a world-renowned electronics manufacturer in Fukuoka, told me:

> It is great that I can show my technological competence to the public outside of the corporate setting. Since it is leisure, I can do whatever I want. However, as public appearances and media coverage became more common, I actually became aware of my responsibility as a robot-builder. Because I don't want my unsatisfactory work to represent the level of Japanese robotics technology publicly, I try to hold to uncompromising standards as a professional engineer in terms of technological quality.

Although he began robot-building as a fun personal activity, Horiwaka now feels that he is caught up in a larger machine as an active representative of Japanese robotics.[13]

In a similar vein Hara, a 39-year-old mechanical engineer engaged in making security robots for a major security company in Japan, observed to me that when getting up on stage with his robot, he identifies himself not only as a robot-builder, but also as a robot entertainer. This reflection is affected by his sense of identity as a Japanese. He comments:

13 Foreign mass media also take part in the formation of these sort of self-reflexive attitudes among robot builders. For instance, when being interviewed on German public television, I found that I was expected to describe my experience, passion, and motivation as a robot builder in a highly simplified manner in response to interview questions based on the underlying assumption that 'the Japanese people have much "closer" relationship with the robot than we do'.

> Because my robot is shown on the stage, I try to design it as friendly-looking as possible...Because I am Japanese, I don't want to build a robot which might make the audience think of the military, especially given the possible influence on children. I had never thought of myself as part of such a big picture until I starting robot-building. My only wish is simple; I just want to make robots that can bring laughter and happiness to people.

Hara's robot is one of the most popular robots in *Robo-One* and appeared onstage at the Aichi World Expo 2006 and on several Japanese television programmes. In September 2008, he was invited to Gran Fiesta Japonesa, a festival held in Ecuador by the Japanese Embassy to celebrate the 90th anniversary of diplomatic relations between Japan and Ecuador. In 2009, Brazil's Robotec Fair called him as well. As his robot gained public exposure both in and outside Japan, the anti-militaristic ideology that Hara regards as distinctly Japanese unexpectedly began to inform his robot-building practices. In humanoid robotics qua nation-branding operation, individual pleasures and motivations are transformed into a national resource.

## Conclusion

Humanoid robotics has become a new site for national identity formation in contemporary Japan. The discursive and affective operation of branding Japan as the nation with the most advanced robotic technologies is part of a radically new, twenty-first century relationship between humanity and technology at the same time as it is part of a new way of constructing national identity. While this nation branding operation promotes politico-economic interests, it is also deeply rooted in Japan's negotiation with modernity. Thus it is unsurprising that the ideology that underpins the nation-building operation is in some ways similar to that which inspired the project of technological nation building through industrial robots in the late 1960s. In both cases the pursuit of *wakon wasai* has been considered essential. The key difference between these two programmes of nation building, however, can be understood as a result of today's increasingly globalized world, in which cultural forms, in association with national images, pervade the global marketplace.

Simultaneously, amateur robot builders' importance in this contemporary project shows that nation branding is not simply a top-down operation enacted through government policy. Rather, it occurs simultaneously on multiple levels and, like Japanese robot culture in general, it takes shape through the agency of innumerable actors, including local governments, the private sector, and individual practitioners who join the fray with a variety of motivations and desires. In this sense, the role of the government is not so much to be a leader in technological innovation, but to put a 'made-in-Japan' seal on robot-building related cultural practices, in hopes that such activities become synonymous with Japanese nationality. This new mode of identity formation is not an expression of the commodification of culture, but of the culturalization of commodities.

## References


Allen, I. (creator). (1965–1968). *Lost in Space*. CBS.

Anholt, S. (2003). *Brand New Justice: The Upside of Global Branding*. Oxford: Butterworth-Heinemann.

Arai, H. (2001). '*Jissen Ni Okeru Robotics Ron*'. *Dai 6 Kai Robotics Symposia Yokou Shu*: 147–53. <http://staff.aist.go.jp/h.arai/robotics/robsym01.html> accessed 24 September 2013.

Aronczyk, M. (2008). '"Living the Brand": Nationality, Globality, and the Identity Strategies of Nation Branding Consultants.' *International Journal of Communication* 2: 41–65.

Asimov, I. (1942). 'Runaround'. In *I, Robot*, pp. 35–51. London: Grafton.

Azuma, H. (2001). *Dōbutsukasuru Posutomodan*. Tokyo: Kōdansha.

Befu, H. (2001). *Hegemony of Homogeneity: An Anthropological Analysis of Nihonjinron*. Melbourne: Trans Pacific Press.

Cameron, J. (1984). *The Terminator*. Hemdale Film.

*Chitekizaisan Suishin Keikaku* (Intellectual Property Strategic Program). (2006). <http://www.kantei.go.jp/jp/singi/titeki2/keikaku2006_e.pdf> accessed 24 September 2013.

Crichton, M. (1992). *Rising Sun*. New York: Alfred A. Knopf.

Fujimura, J. H. (2003). 'Future Imaginaries: Genome Scientists as Sociocultural Entrepreneurs'. In A. Goodman, D. Heath, and M. S. Lindee (eds), *Genetic Nature/Culture: Anthropology and Science Beyond the Two-Culture Divide*, pp. 176–99. Berkeley, CA: University of California Press.

Gibson, W. (1984). *Neuromancer*. New York: Ace Books.

Iida, Y. (2002). *Rethinking Identity in Modern Japan*. New York: Routledge.

Inoue, H. (1993). *Nihon Robotto Sōseiki* 1920–1938. Tokyo: Shūeisha.

Iwabuchi, K. (2002). *Recentering Globalization: Popular Culture and Japanese Transnationalism*. Durham: Duke University Press.

—— (2007). *Bunka no Taiwaryoku*. Tokyo: Nihonkeizaishimbun Shuppansha.

—— (2010). 'Undoing Inter-national Fandom in the Age of Brand Nationalism'. *Mechademia* 5 (Fanthropologies), 87–96.

*Kagakugijutsu Kihon Keikaku* (The Science and Technology Basic Plan). 2006. <http://www.mext.go.jp/english/whitepaper/1302764.htm> accessed 24 September 2013.

Kamoshida, H. (2005). *Robot Gyōkai Saizensen no 28 Nin Ga Kataru! Robot No Genzai To Mirai*. Tokyo: X-media.

Katsuno, H. (2012). 'Robot Dreams: Play, Escape and Masculine-Romanticism in Japanese Techno-Culture'. *PAN-JAPAN*, 8: 104–27.

Kaufman, P. (1993). *Rising Sun*. Twentieth Century Fox.

Low, M. (2003). 'Displaying the Future: Techno-Nationalism and the Rise of the Consumer in Post-war Japan.' *History and Technology*, 19: 197–209.

Lucas, G. (1977). *Star Wars*. Lucasfilm.

McGray, D. (2002). 'Japan's Gross National Cool'. *Foreign Policy*. 130: 44–54.

Martinez, D. P. (2005). 'On the 'Nature' of Japanese Culture, or is there a Japanese Sense of Nature?' In Jennifer Robertson (ed.), *A Companion to the Anthropology of Japan*, pp. 185–201. Oxford: Blackwell.

Massumi, B. (2002). *Parables for the Virtual: Movement, Affect, Sensation*. London: Duke University Press.

Mono Magazine (2006). *'Robotto-shugi Nippon'*. *Mono Magazine* 546: 36–66. Tokyo.

Morley, D. and Robins, K. (199). *Spaces of Identity: Global Media, Electronic Landscapes and Cultural Boundaries*. London: Routledge.

Nelson, V. (2001). *The Secret Life of Puppets*. Cambridge, MA: Harvard University Press.

Olins, W. (2002). 'Branding the Nation: The Historical Context'. *The Journal of Brand Management* 4: 4–5.

Oshii, M. (1995). *Kōkaku Kidōtai*. Bandai Visual Company.

Sato, K. (2004). 'How Information Technology has (not) Changed Feminism and Japanism: Cyberpunk in the Japanese Context'. *Comparative Literature Studies*, 41: 335–55.

Schodt, F. (1988). *Inside the Robot Kingdom: Japan, Mechatronics, and the Coming Robotopia*. Tokyo: Kodansha International.

Segel, H. B. (1995). *Pinocchio's Progeny*. Baltimore, MD: The Johns Hopkins University Press.

Sena, Hideaki. (2001). *Robotto Nijūisseiki*. Tokyo: Bungei Shunju.

Suzuki, K. (2007). *Dai Robottohaku*, exhibition catalogue, 23 October 2007 to 27 January 2008, National Museum of Nature and Science, Tokyo.

Tezuka, O. (1963–66) *Tetsuwan Atomu*. Fuji TV.

Tomino, Y. (1979–80). *Kidō Senshi Gandamu*. Nagoya Broadcasting Network.

Ueno, T. (2002). 'Japanimetion and Techno-Orientalism'. In *The Uncanny: Experiments in Cyborg Culture*, pp. 223–31. Vancouver: Arsenal Pulp Press.

Umetani, Y. (2005). *Robot no Kenkyūsha wa Gendai no Karakurishi ka*? Tokyo: Ohmsha.

Vogel, E. (1980). *Japan as Number One: Lessons for America*. New York: Harper Colophon.

Wilcox. F. M. (1956). *Forbidden Planet*. MGM.

Wood, G. (2003). *Living Dolls*. London: Faber and Faber.

Yonemura, M. (2004). '*Atomu Ideorogii*'. In N. Baba (ed.), *Robotto no Bunkashi*, pp. 74–105. Tokyo: Shinwasha.

Yoshimi, S. (1999). '"Made-in-Japan": The Cultural Politics of Home Electronics in Post-war Japan.' *Media, Culture, and Society*, 21: 149–72.

Yoshino, K. (1992). *Cultural Nationalism in Contemporary Japan*. New York: Routledge.

GRISELDIS KIRSCH, WOLFRAM MANZENREITER AND JOHN HORNE, MERRY WHITE, HIROFUMI KATSUNO AND DOLORES P. MARTINEZ

# Afterword: Reassembling after 3/11

The papers in this volume were initially written before the earthquake and nuclear disaster of the 11th of March 2011, however the book has taken too long, as academic tomes are sometimes wont, to be completed and events have somewhat overtaken us. In light of this and given our exploration of how Japan both has been assembled and how it assembles itself, we could not ignore the events of 3/11 and their aftermath. Below are some of the authors' thoughts on how Japan is still assembling itself and is assembled.

## The Facts

- Japan needs to import about 84 per cent of its energy requirements.
- Its first commercial nuclear power reactor began operating in mid-1966, and nuclear energy has been a national strategic priority since 1973. This came under review following the 2011 Fukushima accident but has been confirmed.
- The country's 50+ main reactors have provided some 30 per cent of the country's electricity and this was expected to increase to at least 40 per cent by 2017. The prospect now is for at least half of this, from a depleted fleet.
- Currently 43 reactors are operable and potentially able to restart, and 24 of these are in the process of restart approvals.

Despite being the only country to have suffered the devastating effects of nuclear weapons in wartime, with over 100,000 deaths, Japan embraced the peaceful use of nuclear technology to provide a substantial portion of its electricity. However, following the tsunami which killed 19,000 people and which triggered the Fukushima nuclear accident (which killed no-one), public sentiment shifted markedly so that there were wide public protests calling for nuclear power to be abandoned. The balance between this populist sentiment and the continuation of reliable and affordable electricity supplies is being worked out politically.

(World Nuclear Association, July 2015)

## An Assemblage of Observations

As an anthropologist raised in North America with fairly 'green' views, Martinez writes, she assumed that Japan was the one modern society which had qualms about relying on nuclear power. Imagine the surprise in 1983 when, while visiting the family of a Japanese friend in the Tohoku region, she was offered a tour of the local power plant as a treat. Such tours were common, part of the public relations that nuclear plants engaged in alongside the practice of hiring locals from rural areas that were at danger of depopulation. The tour emphasized the efficiency and safety features of the plant, especially (it seemed to Martinez as she looked at the displays with a couple in their sixties) designed to calm the fears of the citizens of the only country that had had nuclear bombs dropped on two of its cities. A year and half later in 1985 Martinez toured another plant in Mie Prefecture with the fishermen and diving women of Kuzaki village where she was doing her fieldwork. There the emphasis was on how safe the water washing out of the plant into the ocean was: large koi swam in tanks through which this water flowed. This public relations assemblage of facts was, of course, calculated to calm the fears of what radiation might be doing to the sea which was the basis of the fisherfolk's livelihood. In short, Martinez notes that, she was left – as was most of the Japanese population – with the impression that no nuclear industry was as technologically advanced or as well regulated as that of Japan.

The events of 3/11 have brought us a different Japan, or have they? asks Kirsch. In the wake of the natural disaster, the western media highlighted the 'national character' of the Japanese, which was the supposed reason for their perceived calmness and stoicism in the face of the earthquake. Almost every foreign expert on Japan was asked to comment on this 'unique' endurance, but only if they said what the media wanted. A pertinent example comes from Merry White who was in London during the first few days after 3/11. Asked to comment on the disaster for the American Fox Network, she was incongruously situated in front of a camera in Westminster, with live feed to New York. As the newscaster questioned her, White found herself stymied by the American priorities. On a split screen, she saw scenes of other disasters scrolling down: Hurricane Katrina in New Orleans, an earthquake scene in Haiti. These scenes were of Haitians and African Americans, ostensibly grabbing what they could from destroyed shops. The newscaster prompted: Americans above all want to know, he said, why there is no looting in Japan? White said: 'That is not the highest priority question of this disaster', and went on to describe cooperative humane efforts. Her side of the screen soon went black and the newscaster said: 'Before the broadcast, White said to our producer that the reason there was no looting is that Japan is a homogeneous society ...' Needless to say, White was astounded, furious and ready to sue. This event confirmed for her the essentializing, racist underpinnings of the lowest level of American views of Japan and of its own population.

However, White also notes that Tohoku residents themselves pushed for a return to normalcy as soon as possible. One of the first demands of local citizens in the reconstruction of the devastated tsunami region in Northeast Japan was that coffee houses be built. These, rather than the prefabricated community centres on the emergency housing estates, were seen as a return to 'normal' social institutions and life. Far from the innovation, exoticism and novelty represented by the cafes of modernizing Japan, the cafés of northeast Japan represent comfort, home and the provision of a necessity, both in the beverage and in the social space: a return to the safety of familiar Japan.

Manzenreiter and Horne observe that the impact of the Tohoku earthquake again propelled the development of a different sort of community discourse and involvement in the region and Japan. Football played its part in this, since it had already helped to develop a national community approach to dealing in practical, technical, political and emotional ways with the disaster. A large number of stadiums were damaged, the most serious being at Sendai, Kashima and J2 side Mito Hollyhock. The J-League suspended its regular season for six weeks after the earthquake and the league and clubs throughout the nation assisted in various ways – holding charity matches and other fund-raising events. J-League players volunteered to help distribute emergency supplies and to organize soccer schools for children orphaned and those left homeless. Fan clubs became engaged in raising donations and did other volunteer work; all-star teams played against a side named Team as One, staffed with J-League greats and players from abroad.

When the J-League resumed operations in late April, stadiums all over the country were decorated with banners expressing compassion and the commitment to rebuild Japan. Kashima Antlers players raised a banner after their AFC Champions League match at Suwon in South Korea which read 'With Hope We Can Cope'. According to the J-League Newsletter (May 2011, 47: 8), messages 'of sympathy and support for Japan had been flooding in from the global football community ever since the disaster struck'. Former J-League players showed their support for the victims of the earthquake in various ways and former Japan head coach Zico played in a charity football match in Curitiba in Brazil. Before the UEFA Champions League match between Real Madrid and Olympique Lyonnais a banner was unfurled that read (in Japanese): 'We are together with everyone in Japan.' Another Spanish club, Valencia, wore shirts with their names written in *katakana* script for one match.

These developments prompted Mike Plastow, Japan correspondent of the magazine *World Soccer*, to write:

> The J-League was always intended to be about community. It has always been about supporting each other through sport; about engaging the whole community so that every member can grow individually and together. This was the original vision. This year, when Japan needed it most, the football community coped. It provided a wonderful focus for hope.
>
> (*J-League News Letter* 2011, 47: 8)

This community spirit, linking football via the national team with the entire nation was exemplified by the collective joy about the success of the women's football team later in 2011, when they won the Women's World Cup. However, the ultimate Cinderella story was provided by the team from Sendai, the J-League hometown closest to the epicentre of the magnitude 9.0 earthquake. Prior to 2011, Vegalta Sendai had played only three seasons in the first division and hardly qualified as a contender for the J-League title. Yet in April 2011, with reconstruction work still going on in the badly hit home stadium, fans unfurled a giant banner during an away match that boldly promised 'No defeat until we regain our hometown' (*furusato o torimodosu made oretachi ga makenai*). This was the start to a remarkable series of wins and draws that ultimately earned Sendai the fourth rank and in the following year even the runner's up position.

However, Kirsch reminds us, the world and the Japanese population remained shaken by the fact that a nation as technologically advanced as Japan which had been able to deal with a natural disaster and start to rebuild had not been able to contain the nuclear meltdown in the Fukushima Dai'ichi power plant complex. Incredulously the world observed hydrogen explosions in two of the crippled reactors that spread radioactivity in the region. All images of Japan as a hyper-technologized nation that had persisted since Sony invented the Walkman in the 1980s were either shattered or seriously indented.

Thus Japan became victim to nuclear fallout for a third time in its modern history. This time, however, it led to the displacement of hundreds of thousands of people, as they had to evacuate the rapidly drawn exclusion zone with its high levels of radiation, a zone which seemed to change its boundaries over time, displacing more people. The news quickly spread that these evacuees were bullied in the rest of Japan and not being served at gas stations if they had a Fukushima number plate; again the media called upon *nihonjinron* tropes to explain this phenomenon by the alleged Japanese tendency to eradicate 'difference'.

The western media were soon constructing contrasting images of Japan, calm and supportive versus ostracizing and exclusive. For better or for worse, these representations reveal that *nihonjinron* as a means of assembling Japan are far from dead, whether created by the Japanese or foreigners. In Japan itself, the slogans *Gambarō Nippon* or *Gambarō Tōhoku* (hold on, Japan/Tōhoku) were coined to forge national unity while the word *kizuna* (bonds) became the word of the year. The almost mythical qualities of these words were thus summoned to stress the 'bonds' between all of the Japanese, making the discourse nationalistic and slightly reminiscent of wartime morale boosters. With Shinzō Abe returning to the office of Prime Minister in 2012, after the short interregnum in which the Democratic Party of Japan managed to change little and then got caught up in the crisis, the national policy has turned towards national self-assertion. Abe's re-election in 2014, with another low voter turn-out,[1] appears to point to the continued disenchantment of Japanese voters with all the mainstream parties. What the large portion of non-voters (47.6 per cent) wants remains unclear as many citizens seem unable to find a party that represents them.

Academically, the triple disaster has also led to new research on Japan. We might even go as far as to postulate Japanese Studies as falling into pre- and post-3/11 categories, as Gerteis and George (2014) argue. The effects of nuclear disaster, the research on the 'safety myths' and political implications of the *genpatsu mura* (nuclear village) that became synonymous in the media with the 'cosy relationship' between politics and energy providers which had helped to perpetuated this myth, became sudden buzzwords in Japanese Studies. At the same time, anthropological research on how the victims of the triple disaster have coped is being done (cf. Gill et. al. 2015), while literature on, theatre performances about and the photography of the invisible nuclear contamination have become topics of research and

1 Currently estimated at about 52 per cent as opposed to Japan's usual voter turnout of 71 per cent.

attention as have films, television dramas and documentaries on post-3/11 Japan.

However little of this research is being done from within Japan – we seem to have a foreign academia-driven discourse on how Japan has changed when life in Japan continues as normal for most Japanese who seem to be taking a conservative approach to possible social change – and the nuclear disaster has been all but forgotten. Thus research on Fukushima continues and flourishes, creating both a scholarly and subjective disjuncture between foreigners and most natives on the topic.

Yet the image of Cool Japan continues to thrive – perhaps in a form different from that which creators had expected it to be, as much of Cool Japan is consumed in murkier channels on the Internet. Nonetheless, Abe and his policy makers seem to be determined to control Japan's image in the rest of the world – and successfully so, as winning the bid for the 2020 Olympic Games in Tokyo might indicate. These representations of Japan as cool widen once more the gap between Japan and the rest. Take, for example, NHK World: an English-language service, it is broadcast to 270 million households around the world, but unlike similar worldwide news services such as CNN or BBC World, NHK has a clear agenda to educate the world about Japan and broadcasts a great deal of programmes that underpin Japan's uniqueness within the global context.

All the while Japan remains as enmeshed within global flows as ever and its trade relations with the rest of the world – particularly the United States and China – flourish, leading to an interesting new shift in the East Asian power balance, again skewing traditional centre–periphery models. Territorial frictions and ever increasing nationalistic discourses throughout East Asia complicate the contemporary political scene. The necessity of constructing an 'us' versus 'them' discourse has moved to the level of policy making; equating the need to maintain national unity in the face of nuclear disaster with that of being united in front of a common enemy. Japan is thus continuously being assembled on the global and regional stage through various actors – and with various actants. In brief, while much has become different in Japan, much has remained the same in the perceptions of and the study of Japan.

Katsuno sums up the situation well when he argues that while in the aftermath of the tragic Fukushima Dai'ichi nuclear disaster of 2011, both the mass and social media were flooded with waves of doubt regarding the legitimacy of modern science and technology; now with hindsight, these criticisms have ended up seeming rather superficial. By no means is he arguing for the modern progressive myth that science and technology inexorably lead to a better future. However, in spite of the ongoing crisis at the damaged nuclear power plant, politico-economic actors in Japan continue to make strategic and aggressive use of this myth to promote their agendas (cf. Samuels 2013) and the myth of progress retains a certain rhetorical power in Japanese contemporary society. Although some intellectuals have long argued that the postmodern condition de-legitimates 'grand narratives', this unprecedented technological disaster ironically revealed the enduring strength of the enchantment of the Japanese nation with science and technology.

## References

Gerteis, C. and George, T. S. (2014). 'Beyond the Bubble, Beyond Fukushima: Reconsidering the History of Postwar Japan'. In *The Asia-Pacific Journal*, 12(8/3). <http://www.japanfocus.org/-Christopher-Gerteis/4080>.

Gill, T., Steger, B. and Slater, D. H. (2015). *Japan Copes with Calamity*, 2nd rev. edn. Oxford: Peter Lang.

J-League. (2011). *J-League News: Official Newsletter* 47. Tokyo: Japan Professional Football League.

Samuels, R. (2013). *3.11: Disaster and Change in Japan*. Ithaca, NY: Cornell University Press.

World Nuclear Association. (updated July 2015). 'Nuclear Power in Japan'. <http://www.world-nuclear.org/info/Country-Profiles/Countries-G-N/Japan/>.

# Notes on Contributors

JOY HENDRY is Professor Emerita of Anthropology at Oxford Brookes University. She has worked and published on Japanese society: family, marriage, child-rearing, education, mathematics, politeness and, more generally, cultural display, Indigenous museums and culture centres, theme parks, global networks and Indigenous science. Her most recent book is *Science and Sustainability* (Palgrave, 2014).

JOHN HORNE is Professor of Sport and Sociology at the University of Central Lancashire. He is the author, co-author, editor and co-editor of numerous books, edited collections, journal articles and book chapters on sport and leisure in society. Recent books include *Sport and Social Movements* (Bloomsbury Academic, 2014) and *Leisure, Culture and the Olympic Games* (Routledge, 2014).

HIROFUMI KATSUNO is Associate Professor in the Department of Human Sciences at Osaka University of Economics, Japan. His current research interests include the socio-cultural impact of new media technologies, the relationships between men/masculinities and technologies, and modern ruins and industrial heritage in Japan.

GRISELDIS KIRSCH is Lecturer in Contemporary Japanese Culture at SOAS, London University. Her publications have focused on representations of China and Korea in the Japanese media. In 2015, her book *Contemporary Sino-Japanese Relations on Screen, 1989–2005, A History* was published by Bloomsbury Academic.

WOLFRAM MANZENREITER is Professor of Social Science Research on Japan at the University of Vienna. He has worked and published extensively on sports in Japan with a focus on football. He has also written on gambling and migration. His recent books are *Sport and Body Politics in Japan* (Routledge, 2014) and *Migration and Development* (Promedia, 2014, in German).

DOLORES P. MARTINEZ is Reader Emeritus in Anthropology at SOAS, University of London, and a Research Associate at the University of Oxford. She has written on maritime anthropology, tourism, religion, gender and popular culture in Japan as well as on women's football in the United States. Her latest publications include *Remaking Kurosawa* (Palgrave, 2009) and *Gender and Japanese Society* (Routledge, 2014).

BRUCE WHITE is Associate Professor at Doshita University. His recent Japan-based research has examined Japanese reggae music, indigenous Ainu, Japanese generational change and global identities, civil participation, the Buraku underclass, and the role that taxis play in spreading local information and identity. His book *Japan's Changing Generations* was published in 2012 by Routledge.

MERRY WHITE is Professor of Anthropology at Boston University and has published extensively on Japan, in recognition of which she was awarded the Order of the Rising Sun by the Japan government in 2013. She also has received the Japan's Society's John E. Thayer award for her contributions to US-Japan relations. Her latest book, *Coffee Life in Japan*, was published in 2012 by the University of California Press and was recognised by the Association for the Study of Food and Society as one of the best two publications of 2013.

HEUNG-WAH WONG is Associate Professor at the University of Hong Kong. His research interests include the anthropology of business, the globalization of Japanese popular culture, western social thought and the comparative study of Japan and China. He has recently published *Japanese Adult Videos in Taiwan* (with Hoi-yan Yau; Routledge, 2014).

CHRISTINE R. YANO is Professor of Anthropology at the University of Hawai'i at Manoa. Her interests focus on Japan, and more recently Japanese Americans, and the processes by which nation-cultures construct and sustain themselves, in particular through forms of popular culture. Her most recent book is *Pink Globalization* published by Durham University Press in 2013.

HOI-YAN YAU is a Senior Lecturer at Lingnan University, Hong Kong. Her research has been on the globalization of Japanese pop culture, the comparative study of pornography, sexual cultures and gender politics in East Asia. Her most recent publications include 'Cover Versions in Hong Kong and Japan: Reflections on Music Authenticity', *The Journal of Comparative Asian Development* 11(2), 320–48 and *Japanese Adult Videos in Taiwan* (with Heung-wah Wong; Routledge, 2014).

# Index